DUE TO THE WEATHER

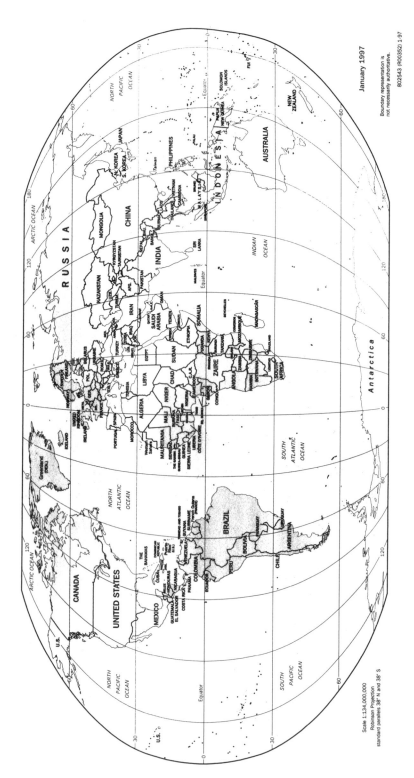

Political Divisions of the World, 1997. (Courtesy of the Library of Congress.)

January 1997

Boundary representation is
not necessarily authoritative.

802543 (R00352) 1-97

Scale 1:134,000,000
Robinson Projection
standard parallels 38° N and 38° S

DUE TO THE
WEATHER

Ways the Elements Affect Our Lives

ABRAHAM RESNICK

Greenwood Press
Westport, Connecticut • London

Library of Congress Cataloging-in-Publication Data

Resnick, Abraham.
    Due to the weather : ways the elements affect our lives / Abraham Resnick.
      p. cm.
    Includes bibliographical references and index.
    ISBN 0–313–31344–X (alk. paper)
    1. Human beings—Effect of climate on. 2. Weather. I. Title.
GF71.R47 2000
    304.2'5—dc21   99–088484

British Library Cataloguing in Publication Data is available.

Library of Congress Catalog Card Number: 99–088484
ISBN: 0–313–31344–X

First published in 2000

Greenwood Press, 88 Post Road West, Westport, CT 06881
An imprint of Greenwood Publishing Group, Inc.
www.greenwood.com

Printed in the United States of America

The paper used in this book complies with the
Permanent Paper Standard issued by the National
Information Standards Organization (Z39.48–1984).

10 9 8 7 6 5 4 3 2 1

Copyright Acknowledgment

The author and publisher gratefully acknowledge permission to reprint the climographs
found at the end of each chapter. They are from Gilbert Tanner's *A Collection of Selected
Climographs* (Eau Claire, Wis.: The Eau Claire State Cartographic Institute, 1964). Copyright
© 1964 by Gilbert Tanner. Reprinted with permission.

For my sister Edythe Resnick Seidenberg
Weathered many a storm
Hardly weathered

With credit and gratitude to Professor Gilbert Tanner of Eau Claire, Wisconsin, for permission to use an array of his *Selected Climographs*, which appear in this book.

A special thank you to Al Schmidt for his review and critique of the manuscript and to Nanette Voorhees for her invaluable assistance in helping with the production of the book.

For Gilda, a much deserved tribute for her unique ability to dissipate many overcast days in order for *Due to the Weather* to find its way under clear skies.

CONTENTS

Contents

INTRODUCTION

Climate is generally recognized as the most important component of the natural environment. It provides us with major essentials for living and greatly influences how we function and many aspects of societal interactions. The interactions between man and climate can be both favorable and unfavorable. Understandings of the relationships between man and climate and the causes and effects of weather and climate are most beneficial in planning for, adjusting to, and coping with most kinds of climatic phenomena. It is apparent, therefore, that how and why man lives as he does in various parts of the world is "Due to the Weather."

There is an often repeated saying that "climate is what you expect, but weather is what you get." Throughout this book, however, the terms weather and climate are used interchangeably. By definition weather is the state of the atmosphere at any given time and place. The conditions of the temperature, pressure, wind, humidity, cloudiness, and precipitation are all basic elements of weather. When they occur over a short period of time, perhaps for several hours or days, they are deemed to be weather. Climate, on the other hand, is regarded as the average statement of the weather conditions of a place or region over an extended period of time. At least twenty to thirty years of data collection and synthesis are necessary to generalize or draw climatic conclusions.

Wherever man lives, he is dependent on a favorable climate to provide him with the essentials of living. They include good air, water, food, and shelter. Periodically, weather conditions can develop that are unfavorable, even violent, and that can lead to distress, disaster, or death. *Due to the Weather* reinforces by examples how people are greatly impacted by both positive and negative aspects of weather. It also illustrates how weather influences, or in some instance determines, the lifestyles of people worldwide. Geographers and other social scientists recognize the major role weather and climate play in shaping human behavior and cultural activities.

Many of the themes and topics in this book identify specific ways the elements can determine the clothing people wear, their leisure time pursuits, customs, history, economic and social relationships, and various transportation modes. Weather and climate, it is noted, are factors in the spread of disease, the state of people's health, and the manner in which those subject to extraordinary weather occurrences react to others during times of crises.

People tend to take so-called normal weather in stride and infrequently refer to it in their discussions and daily greetings. Exceptionally pleasant weather can, however, quickly become a topic of conversation for those who delight in observing fine weather. In contrast, the arrival of a severe storm or other harsh or inclement weather is usually the source of great concern and even fear that the forbidding weather might cause property damage, human injuries, or even death. When that occurs, it seems that everyone wants to talk about it, eagerly giving advice and warnings on ways to deal with its severity.

This book underscores a number of advances being made by modern technology, particularly those used by weather forecasting centers designed to alert possible victims who may be in harm's way and to temper future weather catastrophes.

Due to the Weather endeavors to provide the reader with an appreciation and awareness of how weather and climate factors play a significant role in determining many human relationships, past and present. Moreover, it is hoped that this book will help foster a greater knowledge and understanding of certain weather conditions found throughout the world. It is also a desire that the content covered may enable the reader to have opportunities to consider the extent that the weather factor may have as a major determiner in shaping people's lives and destinies.

The volume reveals a broad array of meaningful, diverse, and rather startling ways the elements affect our lives. It presents hundreds of meteorological events, expository facts, and anecdotes that demonstrate the diverse methods people utilize in coping with or adjusting to weather phenomenon.

This book is divided into two parts. The first describes in detail twenty different kinds of weather elements and specific ways each impacts people's lives. The second part identifies ten distinct topics that best illustrate how weather affects human culture and activities.

CLIMOGRAPHS AND SUMMARIES

At the end of each chapter four representative weather station climographs are provided for analysis. They graphically reflect the general climatic conditions associated with worldwide places that most often experience the type of weather described in that chapter. The climograph selections are generally places that are referred to in the text or are located within the region mentioned. A number of the climographs correspond with the various topics treated.

The 120 climographs presented offer monthly records of rainfall and temperature data, as well as locational and elevation information. In addition, each set of climographs is accompanied by a brief summary that serves to highlight significant facts, concepts, and understandings that reinforce ways weather interacts with people and affects many human responses and activities.

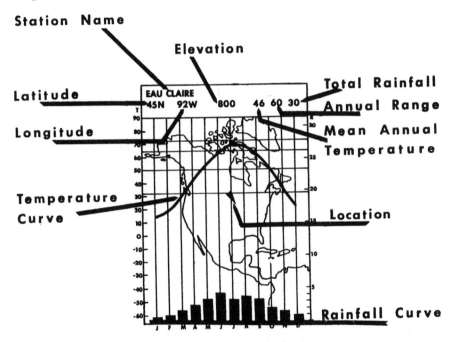

Each climograph presents data as represented in this key.

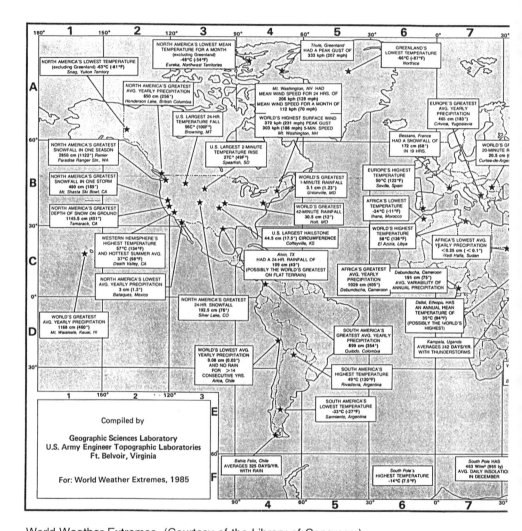

World Weather Extremes. (Courtesy of the Library of Congress.)

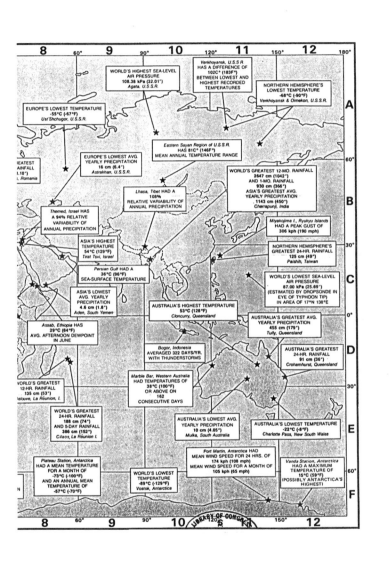

Map labels:

8 60° **9** 90° **10** 120° **11** 150° **12** 180°

WORLD'S HIGHEST SEA-LEVEL AIR PRESSURE 108.38 kPa (32.01") Agata, U.S.S.R.

Verkhoyansk, U.S.S.R. HAS A DIFFERENCE OF 102C° (183F°) BETWEEN LOWEST AND HIGHEST RECORDED TEMPERATURES

NORTHERN HEMISPHERE'S LOWEST TEMPERATURE -68°C (-90°F) Verkhoyansk & Oimekon, U.S.S.R.

A

EUROPE'S LOWEST TEMPERATURE -55°C (-67°F) Ust'Shchugor, U.S.S.R.

Eastern Sayan Region of U.S.S.R. HAS 81C° (146F°) MEAN ANNUAL TEMPERATURE RANGE

60°

EUROPE'S LOWEST AVG. YEARLY PRECIPITATION 16 cm (6.4") Astrakhan, U.S.S.R.

REATEST RAINFALL 1.10") , Romania

WORLD'S GREATEST 12-MO. RAINFALL 2647 cm (1042") AND 1-MO. RAINFALL 930 cm (366") ASIA'S GREATEST AVG. YEARLY PRECIPITATION 1143 cm (450") Cherrapunji, India

B

Lhasa, Tibet HAD A 108% RELATIVE VARIABILITY OF ANNUAL PRECIPITATION

Themed, Israel HAS A 94% RELATIVE VARIABILITY OF ANNUAL PRECIPITATION

Miyakojima I., Ryukyu Islands HAD A PEAK GUST OF 306 kph (190 mph)

ASIA'S HIGHEST TEMPERATURE 54°C (129°F) Tirat Tsvi, Israel

NORTHERN HEMISPHERE'S GREATEST 24-HR. RAINFALL 125 cm (49") Paishih, Taiwan

30°

Persian Gulf HAD A 36°C (96°F) SEA-SURFACE TEMPERATURE

WORLD'S LOWEST SEA-LEVEL AIR PRESSURE 87.00 kPa (25.69") (ESTIMATED BY DROPSONDE IN EYE OF TYPHOON TIP) IN AREA OF 17°N 136°E

C

ASIA'S LOWEST AVG. YEARLY PRECIPITATION 4.6 cm (1.8") Aden, South Yemen

AUSTRALIA'S HIGHEST TEMPERATURE 53°C (128°F) Cloncurry, Queensland

AUSTRALIA'S GREATEST AVG. YEARLY PRECIPITATION 455 cm (179") Tully, Queensland

0°

Assab, Ethiopia HAS 29°C (84°F) AVG. AFTERNOON DEWPOINT IN JUNE

Bogor, Indonesia AVERAGED 322 DAYS/YR. WITH THUNDERSTORMS

AUSTRALIA'S GREATEST 24-HR. RAINFALL 91 cm (36") Crohamhurst, Queensland

D

WORLD'S GREATEST 12-HR. RAINFALL 135 cm (53") lelouve, La Réunion, I.

Marble Bar, Western Australia HAD TEMPERATURES OF 38°C (100°F) OR ABOVE ON 162 CONSECUTIVE DAYS

30°

WORLD'S GREATEST 24-HR. RAINFALL 188 cm (74") AND 5-DAY RAINFALL 386 cm (152") Cilaos, La Réunion I.

AUSTRALIA'S LOWEST AVG. YEARLY PRECIPITATION 10 cm (4.05") Mulka, South Australia

AUSTRALIA'S LOWEST TEMPERATURE -22°C (-8°F) Charlotte Pass, New South Wales

E

Plateau Station, Antarctica HAD A MEAN TEMPERATURE FOR A MONTH OF -73°C (-100°F) AND AN ANNUAL MEAN TEMPERATURE OF -57°C (-70°F)

WORLD'S LOWEST TEMPERATURE -89°C (-129°F) Vostok, Antarctica

Port Martin, Antarctica HAD MEAN WIND SPEED FOR 24 HRS. OF 174 kph (108 mph) MEAN WIND SPEED FOR A MONTH OF 105 kph (65 mph)

Vanda Station, Antarctica HAD A MAXIMUM TEMPERATURE OF 15°C (59°F) (POSSIBLY ANTARCTICA'S HIGHEST)

60°

F

8 60° **9** 90° **10** 120° **11** 150° **12**

PART I

DIFFERENT KINDS OF
WEATHER ELEMENTS

A snowmobile sticks out of the rear of a school where an avalanche swept through during a New Year's celebration in Quebec, Canada in 1999. At least nine people were killed. (*Credit:* AP/Wide World Photos. Reprinted with permission.)

1

AVALANCHES

In mountainous regions, great masses of loosened snow, ice, earth, or rocks sometimes slide down the side of a mountain, destroying everything in their path. These slides are called avalanches, from the French word meaning descent. Smaller masses of snow and ice are called snowslides. An avalanche can be triggered by a loud noise such as a blast or thunder.

Avalanches frequently occur in the spring when the snow begins to melt. The most dangerous avalanches are started by strong winter winds, which cause great masses of solid snow to drop with tremendous force toward the valley below. In the summer, ice avalanches break off from high glaciers during periods of thawing and crash to lower elevations.

WHERE AVALANCHES OCCUR

The Alpine region of Europe suffers more avalanches than anywhere else in the world. Deaths related to avalanches in the Alps from 1993 to 1998 totaled 504. France reported the most, with 168 fatalities. Austria had 142 deaths, Switzerland 104, and Italy 90. For a like period of time the entire United States lost 148 victims to avalanches. Unfortunately the average number of Americans killed by avalanches has more than tripled, rising from seven a year in the 1970s to twenty-four a year in the 1990s. This is mainly due to a marked increase in backcountry skiers, snowboarders, and snowmobilers. The rapidly increasing housing, condominium, and ski resort construction in the mountains causes avalanche experts to see a potential threat to human safety as more vacationers tend to move into avalanche-prone areas in the Rockies and other high mountainous parts of the United States and Canada.

HISTORIC AVALANCHES

Avalanches can be very destructive. Entire villages in the Alps have been destroyed by them with considerable loss of human life. Especially vulnerable are mountain climbers, skiers, snowmobilers, and travelers in mountainous terrain.

In 1885 an avalanche at Alta, Utah, killed sixteen and buried thirteen people who were eventually rescued. Near Alta, in the Wasatch Range, avalanches killed 200 people between 1865 and 1915. In 1910 a devastating avalanche plummeted down a mountain in Washington, sending three locomotive engines and some cars over the side, burying them under tons of snow and killing 100 passengers. Heavy snows and high winds have caused avalanches in mountainous regions throughout the world, which have resulted in numerous casualties. The worst avalanche ever at a U.S. ski resort took place near Lake Tahoe, California, on March 31, 1982 when seven people were killed.

RECENT AVALANCHES

An avalanche smashed into a remote Icelandic fishing village in January 1995. Buried in the rubble were sixteen dead, half of them children, who died in their beds. Three days later perhaps as many as 175 people were killed after an avalanche slid down a Himalayan Mountain slope near Jammu, India. During the first weeks of the skiing season of 1997, at least sixteen skiers lost their lives due to avalanches in Switzerland. Hundreds of snow slides swept through British Columbia, Montana, and Utah during early January 1998. Experts constantly monitor huge glaciers in Italy's western Alps, which have been known to have sections drop onto inhabited valleys.

More than thirty-two bodies were pulled from the snow in January 1998 after an avalanche buried vehicles on a mountain highway near Tehran, Iran. On January 25, 1998, the weight of thirty-two students and six adults over the fragile powder of five feet of new snow on an Alpine ridge near Embrun, France, during a snowshoe trek, caused a deadly avalanche killing eleven in the party. Warning flags and safety signs posted in the area that indicated the danger of an avalanche possibility were ignored. A dozen climbers nearly 12,000 feet up Mount Rainier were swept away in June 1998 by an avalanche, precipitated by warm spring weather that caused the snow base to become slushy. Eventually all but one were rescued via airlifts by army helicopters. The worst single accident on Mount Rainier occurred on June 21, 1981, when eleven people died in an icefall. It is estimated that more than 300 hikers and climbers died in accidents on the mountain over the past 100 years.

An avalanche swept down on a gathering on January 1, 1999, at a remote Eskimo village called Kangiqsualujjuag, 950 miles northeast of Montreal, Quebec. Nine residents of the Inuit community were killed and twenty-five others were seriously injured when tons of snow roared down a 250-foot hill only eighteen feet from the school gymnasium, knocking out a wall where more than 400 were celebrating the holiday.

A previous safety report had recommended that steel snow barriers be built around the hill to avert an avalanche thrust, but unfortunately no barriers were built.

When a powerful avalanche struck at the tiny hamlet of Les Houches in the French Alps on February 11, 1999, the villagers there were much more fortunate than the Inuit townspeople. The French, aware of the potential dangers that could result from a massive snowslide in their Alpine environment, built a formidable antiavalanche barrier to fortify their town against being overwhelmed by tons of rapidly falling snow and ice from the lofty mountains overhead. In the nearby valley around Chamonix, only three days previously, the people were not so lucky. There a huge avalanche swamped the area, killing twelve and demolishing a number of unguarded chalets. Chamonix had failed to erect any wall that would withstand a snowslide. It was the worst avalanche in the region in ninety-one years.

Heavy snowstorms throughout France, Switzerland, Austria, and Germany triggered a series of avalanches in that part of Europe during early February 1999. Nineteen people perished. The snowstorm, accompanied by an unrelenting bitter cold of sub-zero temperatures, hampered the rescue operations of courageous and dedicated workers using sensors and trained dogs. Their search mission turned out to be futile. Once again a devastating avalanche had brought human activities to a standstill.

Large masses of snow and ice detached from mountain slopes once again overspread the French Alps and the Pyrenees Mountains bordering Spain only ten days after the earlier avalanches. More than a dozen French tourists were killed or unaccounted for. For the Alps this proved to be the heaviest snowfall in decades with depths recorded at more than eight feet. Many chalets were engulfed or destroyed. Tens of thousands of tourists were trapped in ski resorts as roads and rail links were buried by tons of snow. Frequent helicopter flights had to be employed to airlift hundreds to safety.

Deadly avalanches, some at speeds of 200 miles an hour, also brought big snows to the mountainous regions of the western United States and Canada about the same time. At least seven fatalities were reported. Some were attributed to high risk and careless skiers who ventured off controlled trails into wilderness areas where posted signs warned of potential avalanche dangers. Those that had to be rescued were subject to a minimum fine of $500.

CAUSES AND THEORIES

When a series of heavy snowstorms or avalanches take their tolls, speculation points to the possible weather phenomenon identified as La Niña. It is thought that an unusually cold pool of Pacific Ocean water near the

Equator induces widespread atmospheric changes that can result in extreme weather conditions. For the safe skier, that could mean an abundant long-lasting amount of snow and extended days of enjoyment on the slopes. For others bent on breaking the rules, or foolhardy enough to challenge an avalanche, La Niña could indirectly bring about tragedy.

A heavy snowfall of over twenty feet during a two-week period can pose a danger for sports enthusiasts by triggering an avalanche. Nearly sixty feet of snow caused a fatal avalanche on Mount Baker in southwestern Canada early in the snow season of 1999. An extremely large snow avalanche has the force to uproot mature forests and even to demolish concrete buildings with impact pressures of more than 100 tons per square meter. A dry powder avalanche may reach velocities of sixty-five to eighty-five meters per second. It is estimated that a rapidly accelerating avalanche of 160,000 tons may be the equivalent of a 20 million horsepower force, nearly 3,000 times greater than a locomotive.

The heavy snows in the Alps during January and February 1999 brought devastating avalanches to the region that at some sites measured fifteen to twenty feet deep and 300 to 500 feet wide. During a three-day period at the end of February, thirty-eight ski vacationers lost their lives in and around the western Austrian villages of Galtur and Landek. The series of avalanches were attributed to the almost unequaled amounts of snow and high winds that caused massive weighty drifts. The overload probably gave impetus to the mass of snow, ice, and rocks to slide suddenly and swiftly down the mountains, particularly where the slopes were inclined 25 to 50 degrees. These vast avalanches demolished many chalets and even sturdy concrete Tyroleon mountain houses as well as numerous automobiles.

The six weeks of "wicked winter" in 1999, with snowfalls higher than roof tops, proved to be the worst since a series of snowslides killed sixty-six people in the mountains of northwest Austria in the village of Blons in January 1954.

PRECAUTIONARY MEASURES

In the interest of safety, avalanche controllers now employ a number of precautionary measures in an attempt to prevent large avalanches from causing great harm to people and property. They include utilizing bombs designed to set off small avalanches in order to allow huge snow masses to move down a mountain by breaking up the heavy accumulations into fragments. There is also a campaign to keep ski enthusiasts away from remote, dangerous, unsupervised lands. In those areas, as in Colorado, as many as 20,000 avalanches occur each year. Scientists are currently developing techniques to predict avalanches by analyzing the relationships between meteorological and snow-cover factors.

Avalanche forecasters are currently devising a number of tests and methods to ferret out dangerous weaknesses in snowpacks and snow stability. Skiers, snowmobilers, and other snow sport enthusiasts are being encouraged to take avalanche safety courses in advance. They are also expected to heed the advice of local avalanche center personnel and be alert to short-term weather conditions.

EFFECTS OF AVALANCHES

When avalanches take place, visitors become disinterested in vacationing in the affected areas. To prevent tourist business declines Avalanche Warning Centers undertake safety-orientation programs and classes designed to offer greater security to skiers. The centers broadcast from mountain radio stations and forewarn skiers of possible danger areas. The procedures have proven successful in preventing deaths in approved ski areas.

Despite the numerous attempts at precautionary measures that are presently being employed throughout worldwide ski areas—such as preventing people from entering dangerous areas before explosives are detonated to set off "artificial avalanches"—fatalities continue to occur. In locations where data are recorded (North America, Europe, and Japan) there are typically 150 to 200 avalanche deaths each year.

AVALANCHES

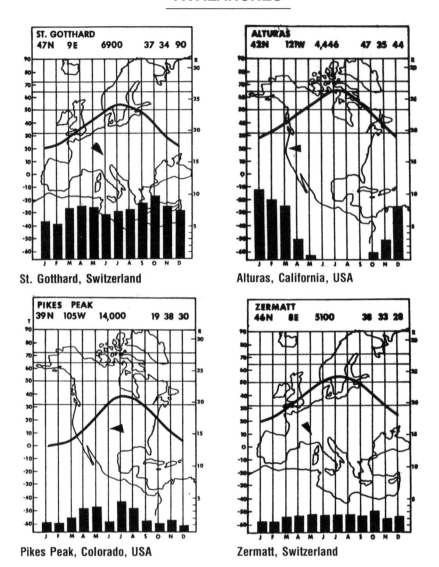

St. Gotthard, Switzerland

Alturas, California, USA

Pikes Peak, Colorado, USA

Zermatt, Switzerland

These high elevation weather stations, which have massive winter season snow accumulations as indicated by the temperature and rainfall curves (see key in the Introduction) and are located adjacent to steep mountain slopes, are places where the potential for avalanches are great. The disastrous snow slides inundate people, cause death and injury, damage property, and can halt all means of travel and communication in the region.

SOURCES

"Avalanches, Blizzards Bury Alps." *New York Times*. February 23, 1999.

Brooke, James. "U.S. Avalanche Experts See Future Threat in Rockies." *New York Times*. February 25, 1999.

"8 Killed in Avalanches in British Columbia." *New York Times*. January 5, 1988.

Gaskell, T. F., and Martin Morris. *World Climate: The Weather, the Environment and Man*. New York: Thames and Hudson, 1979.

Goodman, David. "When the Mountain Falls." *Outside* (Santa Fe, N. Mex.), vol. XXV, no. 2, February 2000.

Higgins, Alexander G. "4 More Avalanche Victims Pulled from Snow in Alps." *Sun-Sentinel* (Fort Lauderdale, Fla.). February 12, 1999.

Ines, J. D., and R. G. Barry. *Arctic and Alpine Environments*. London: Metuchen, 1974.

Jenkins, McKay. "And None Came Back." *Outside* (Santa Fe, N. Mex.), vol. XXV, no. 2, February 2000.

Ludlum, David M. *The Weather Factor*. Boston: Houghton-Mifflin, 1984.

"Mount Rainier Avalanche Strikes 12 Climbers; 4 Are Rescued." *New York Times*. June 12, 1999.

National Disasters of North America (supplement). *National Geographic*. July 1998.

"Nine Dead, 25 Injured in Quebec Avalanche." *New York Times*. January 1, 1999.

Phillips, Jan. "Avalanche Barrier Saves French Town." *Sun-Sentinel* (Fort Lauderdale, Fla.). February 12, 1999.

Schneider, Stephen H., ed. *Encyclopedia of Climate and Weather*. New York: Oxford University Press, 1996.

"16 Feared Dead in Avalanche." *New York Times*. January 17, 1995.

"32 French Teen-age Students and Adult Chaperones Set Off Deadly Slide on Alpine Ridge." *New York Times*. January 25, 1998.

"32 Killed in Avalanche near Tehran." *New York Times*. January 16, 1998.

Verhovek, San Howe. "Big Snows Brings Sportsmen, but Also Deadly Avalanches." *New York Times*. February 16, 1999.

Whitney, Craig R. "After the Avalanches, Europe Skis Into Spring." *New York Times*. March 28, 1999.

———. "As the Lucky Escape the Alps, a Snowslide Claims Others." *New York Times*. February 25, 1999.

An anvil-top cumulus cloud is an indicator of very rapidly moving vertical air currents. These clouds are extremely dangerous for aircraft. Note the nimbus rain clouds at the lower level of the cloud formation. (Courtesy of National Oceanic and Atmospheric Administration.)

2

CLOUDS

A cloud is a visible mass of condensed water vapor, either in the form of tiny drops of water or ice crystals, suspended in the atmosphere. Generally clouds are classified according to their height: high (16,500 to 45,000 feet), middle (6,500 to 23,000 feet), and low (0 to 6,500 feet). The four chief forms of clouds are: cirrus, which are high only white clouds; stratus, which means spread out and most often are long, low, gray layered with a uniform base; cumulus, a Latin word meaning heap or puffy, are white clouds resembling cotton balls or globs of whipped cream; and nimbus, a kind of shapeless dark gray rain cloud, heavy with moisture.

The cirrus clouds generally are predictors of fair weather if they do not thicken. They appear to be light and wispy. Once they lower and merge with other types of cloud formations they thicken, which usually indicates rain or snow will occur in twelve to twenty-four hours. The much lower level stratus clouds are grayish in color and extend in a long layer with a uniform base. They tend to signify fair weather. The dome-shaped cumulus clouds most often foretell fair weather when they stand alone in the sky in a billowy manner. During the summer months, however, when the warm moist air rises rapidly into the atmosphere, they have a tendency to become heavy and flatten by late afternoon. When that takes place they become altocumulus (middle level) or patches of dark stratocumulus type clouds, and they can bring on a thunderstorm or rain before the end of the day.

The most obvious and ominous cloud type of all is the nimbostratus, an extensive, dark low-level mass of gloomy murkiness, easily leading the sky observer to conclude that the "rain clouds" are an imminent threat. The nimbostratus clouds arrive slightly ahead of a warm front and can bring long steady rain or snow in winter.

EFFECTS OF CLOUDS

Clouds are known to have a considerable effect on people's lives. Many individuals, sometimes subconsciously, look to the skies for a perception of their own weather hunches. Throughout the ages men have gazed at the clouds in pensive moments. Others fantasize about their ever changing shapes and see images of people, animals, and objects.

Poets, novelists, artists, and photographers have been inspired aesthetically by cloud formations, shapes, and colors. People can readily be influenced in their moods and psychological attitudes by the types of clouds that persist over their locales for extended periods of time. Overcast, dark clouds can lead to ongoing dispositions of depression, lethargy, and glumness. Conversely, bright cumulus clouds of fine weather can bring about a lively, happy aura and cheerful persona in people.

Low blanketlike clouds allow dust particles to accumulate in the atmosphere that can cause smog and pollutants to disturb healthful breathing and bring on eye and respiratory problems for people. Cloud conditions can determine the selection of apparel and head coverings, especially if one anticipates a beach excursion or outdoor activity, such as tennis or baseball. An extended period of cloudy days can lessen the quality or outcome of a vacation, or the visibility of an automobile tour, especially at scenic overlooks. Low-lying clouds can determine aircraft routes and landings. The landing of the space shuttle *Atlantis* was delayed an extra day due to the thick overcast at Cape Canaveral, Florida, on October 5, 1997.

Clouds have been both friend and foe of military operations on air, land, and sea. They may help determine the outcomes of battles by shielding or by providing clear visibility in sighting targets. They assist in giving cover for landings, troop movement, and bombings.

Late in March and early April 1999, overcast skies hampered NATO warplanes bombarding Serb army forces, airfields, and supply depots. Due to clouds and poor visibility they were unable to target air defenses or blunt Serbia's ground offensive in Kosovo. Scores of allied fighter bombers had to abort their missions and return to their bases, unable to release their laser-guided bombs during the cloudy conditions. The accuracy of bombing improves markedly when the cloud ceiling is at least 25,000 feet and visibility is more than five miles.

IMPORTANCE OF CLOUDS TO WEATHER PEOPLE

Meteorologists and weather observers plot the condition of the sky and its cloud cover on maps and analyze their data for scientific recordings and local weather forecasting. They work together in weather stations throughout the world and periodically report on local or area weather conditions. Observations and reports are made of clouds and timely weather elements such as rain, hail, snow, fog, wind, temperature, ice, flood, visibility, lightning, frost, air pressure, and relative humidity. They measure their data and record the information on synoptic maps that are studied by the meteorologist in making weather forecasts. In addition to basic weather instruments, which include thermometers, psychrometers, weather vanes, anemometers, and barometers, many state-of-the-art computers, radar, and an array of electronic devices are employed to gather precise data. During storms weather planes fly into and or above the atmospheric disturbances in order to learn firsthand information about their intensity and direction. Improvements in weather forecasting are saving lives, preventing injuries, and decreasing damage costs to property from severe storms by alerting people to take precautionary procedures well in advance.

CLOUDIEST PLACES IN THE WORLD

The average annual cloud cover for the entire earth is about 61 percent. The greatest amount of cloud cover is found near the Equator and high latitudes. At coastal locations along the eastern Pacific Ocean and western Atlantic Ocean cloudiness is quite prevalent. Those areas have what is classified as a humid marine (West Coast) climate.

Sitka, Alaska; Seattle, Washington; Bergen, Norway; Brest, France; and Wellington, New Zealand, have climates with a propensity for cloudiness. Prevailing winds carry air from the ocean over the land. The marine influences predominate. Coastal mountain ranges tend to block moist Pacific air masses from rising over them; but the absence of transverse mountains in northwestern Europe allows westerly winds from the Atlantic to carry air masses far inland. Even in central Europe the climate is modified by marine influences, thus many of their cities have overcast skies through much of the year.

EFFECT OF CLOUDS ON PROPERTY SELECTION

Selections of home sites and real estate values can even be influenced by cloud conditions. This is especially noticeable along the Pacific coastline where low stratus type clouds frequently appear on the windward sides of mountains late in the afternoon and evening, only to dissipate with the appearance of the late morning sun the following day. In Washington and Oregon moist air is often trapped in the valleys during the winter for weeks at a time and low clouds persist for long periods. The citizens of Portland, Oregon, have an average of 215 cloudy days each year. The people of Anchorage, Alaska, have the greatest number of cloudy days at 259, which is the nation's greatest number of cloud-covered days in a given year.

The real estate saying of "location, location" becomes especially significant where overcast skies are prevalent. People refrain from selecting home sites where views are obscured by low-lying clouds or in areas where roads are often enveloped by misty, foglike conditions. Many home buyers prefer higher ground, above the clouds where the air is clear and dry and exposure to sunlight is not a rarity.

People have been known to resort to a kind of "micro-climate" survey when deciding on a building site. Their criterion sometimes include a selection on the lee or dry side of a mountain, an inland location, or a place where the wind pattern blows away pockets of moist air. Some are even concerned about the excessive moisture of low clouds as a cause of car rust, sticking windows, and the need for frequent house painting.

Most people seem generally unobservant about their daily sky conditions. Clouds are taken in stride rather innocuously unless they serve to enhance a beautiful sunrise or sunset or portend a dramatic change in the weather. In the latter case cloud formations can prove to be a lifesaver. They might be a drought-ending harbinger or harbor electrical discharges of lightning. In either and all instances clouds should not be "overlooked." They may eventually tell a meaningful story.

CLOUDS

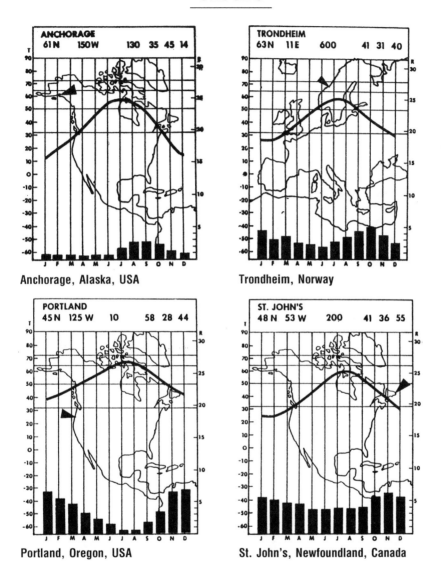

ANCHORAGE
61 N   150 W   130   35   45   14

Anchorage, Alaska, USA

TRONDHEIM
63 N   11 E   600   41   31   40

Trondheim, Norway

PORTLAND
45 N   125 W   10   58   28   44

Portland, Oregon, USA

ST. JOHN'S
48 N   53 W   200   41   36   55

St. John's, Newfoundland, Canada

The low cloud cover in the general area of these cities due to their locations (see key in the Introduction) may extend for weeks on end. The lingering overcast has been known to cause some people to experience depression, lower vitality levels, and a dampened mood as they react to the prolonged gloomy sky conditions.

SOURCES

"Astronaut's Return from the Mir Is Delayed by Cloudy Weather." *New York Times*. October 5, 1998.

Battan, Louis J. *Fundamentals of Meteorology*. Englewood Cliffs, N.J.: Prentice-Hall, 1984.

Fisher, Robert Moore. *How to Know and Predict the Weather*. New York: New American Library, 1955.

Gedzelman, Stanley. "Mysteries in the Clouds." *Weatherwise*. June/July 1995.

International Cloud Atlas. Vol. 2. Geneva, Switzerland: World Meteorological Organization, 1987.

Ludlow, F. H., and R. S. Scorer. *Cloud Study: A Pictorial Guide*. London: John Murray, 1957.

Mason, B. J. *Clouds, Rain and Rainmaking* (2nd ed.). New York: Cambridge University Press, 1975.

National Disasters of North America (supplement). *National Geographic*. July 1998.

Oliver, John E., and Rhodes W. Fairbridge, eds. *The Encyclopedia of Climatology*. New York: Van Nostrand Rheinhold, 1987.

Pearce, E. A. and C. G. Smith. *The World Weather Guide*. London: Hutchinson Publishing Group, 1984.

Rubin, Louis, and Jim Duncan. *The Weather Wizard's Cloud Book*. Chapel Hill, N.C.: Algonquin Books, 1989.

Schneider, Stephen H., ed., "Clouds." *Encyclopedia of Climate and Weather*. Vol. 1. New York: Oxford University Press, 1996.

Scorer, Richard. *Clouds of the World*. Melbourne, Australia: Lothian Publishing, 1972.

Trewortha, Glenn T., and Lyle H. Horn. *An Introduction to Climate* (5th ed.). New York: McGraw-Hill, 1980.

World Meteorological Organization, International Cloud Atlas. Vols. 1 and 2. Geneva: World Meteorological Organization, 1956.

3

DROUGHTS

CAUSES AND THEORIES

A drought is a kind of slow emergency. It never occurs suddenly. A drought results from long-continued dry weather and a lack of sufficient rain, whereby the dryness has become severe enough to cause concern over decreasing supplies of water, crop losses, and forest fire hazards. Droughts usually take place when the normal amount of rainfall of an area is 15 percent below what is usually received for a period of two to four weeks. The lack of rain is often accompanied by high temperatures, dry crumbling earth, hot, dry winds that blow away topsoil, and dried water bodies and wells. Drought causes considerable sickness and death to farm animals and livestock. When a region experiences a prolonged drought, farm products are in short supply, resulting in increased costs to consumers. In areas where farmers grow food only for their own subsistence, a lengthy drought can bring malnutrition, disease, and death due to food shortages. Widespread failure of crops because of a lack of water is one of the primary causes of famine.

The U.S. Weather Bureau distinguishes three categories of drought: (1) absolute drought, a period of at least fifteen days without measurable rainfall, (2) partial drought, a period of at least twenty-nine consecutive days in which the mean daily rainfall does not exceed 0.01 inches, (3) dry spell, a minimum of fifteen consecutive days during which less than 0.04 inches of rainfall is received.

Some geographers and climatologists define conditions as a drought when the annual precipitation is 75 percent below normal or monthly precipitation is 60 percent below normal. Others say that any amount of rainfall less than 85 percent of normal constitutes drought. Drought can be permanent, seasonal, irregular (variable rainfall), and invisible (insufficient rainfall to offset evaporation).

The water on this Georgia ranch is well below the white mark on the standpipe (in the background) which shows where the normal water level should be had there not been a drought there in August 1986. A drought can readily affect animals' size, weight, and health and eventually become a factor in the price of meat. (Courtesy of U.S. Department of Agriculture.)

WORLDWIDE DROUGHTS

Throughout history droughts have inflicted great suffering upon millions of people, worldwide, in the form of hunger, starvation, death, and economic ruin. In drought-stricken regions of Africa observers have reported tens of thousands of victims who have little energy to do daily tasks, mental dullness, and a breakdown in moral standards because of the quest to survive at any cost, resulting in some cases with villagers resorting to stealing food, murder, and cannibalism. Disruption of family life is commonplace under such circumstances. Whole villages are forced to migrate and wander in search of food.

During the periodic droughts in northeastern Brazil, it is not unusual for hungry peasants to rush in to well-to-do towns to loot food stores and government food warehouses. In May 1998 an estimated ten million people in eight states suffered the effects of a six-month drought. During that time 110 looting incidents were reported. Some store owners parceled out free food, and church officials publicly supported the looters who stressed they were compelled to steal food so that their families might survive. (Under Brazilian law, people are permitted to steal food if they find themselves "in a state of necessity.")

During the same six-month drought no rain fell in the Amazon rain-forest. Fires raged throughout the region. Yanomami Indian shamas (the Yanomami's currently are the world's largest existing Stone Age tribe) gathered to perform a sacred ceremony to call down rains. Deep in the jungle the medicine men and elderly mystics known as *xapuris* entered a hallucinogenic trance by snorting powdered bark of a tree and chanted spells to "cool the sun." The practice has been used by the tribal priests for centuries, primarily during extreme droughts or outbreaks of disease. Native Americans have also been known to perform ritualistic rain dances in their pleas for drought relief. In timber-dry Florida after many weeks of desertlike drought in June and July 1998, thousands of church-goers prayed for rain to quell the flames that produced 2,000 wide-ranging wildfires that devoured 450,000 acres of brush and forests. The drought-induced fires forced 100,000 people to evacuate their homes during 1,700 separate fires.

The drought and famine that frequently sweep across Africa are familiar occurrences to most Africans. Over the past four decades, all but five nations south of the Sahara have suffered severe food shortages. An extended period of drought in the 1980s was considered to be one of the worst droughts of the century. The United Nations estimated that twenty-nine million people were at risk of starving during that period. Up to ten million people were on the verge of perishing due to failed rains in Sudan. The far-reaching famine caused the deaths of hundreds of thousands. The victims that survived either fled to displaced persons' camps to be fed by the millions of tons of food supplied by international aid organizations or migrated into grim urban slums. Many suffered terrible impoverishment in their famish-stricken villages, or in despair abandoned their fields and desperately scoured the countryside for any kind of food deemed edible—anything for survival.

EFFECTS OF DROUGHT

Drought can cause famine and untold agony particularly to children in third-world countries and poverty-stricken villages throughout the globe. In North Korea a two-month summer drought in 1997, accompanied by scorching heat, threatened the lives of 60,000 children aged five and under. Their survival was jeopardized due to a severe food shortage. At one school, six- and seven-year-olds were being fed one sparse meal every other day. Children everywhere were bone thin, many too weak to even sit up. Weak, undernourished adults were often much too sickly to provide food or medical relief. During the food scarcity a lot of chronically malnourished and stunted children were abandoned or given up to government orphanages by their parents. Also, in late 1997 a drought in Indonesia forced nearly 82,000 village people in at least 130 villages to roam the forests in search of wild yams after their crops failed. More than 600 villagers in the western half of the Pacific island of New Guinea succumbed to disease from drinking unsafe water during the drought.

Below-average rainfall amounts, even in more advance countries, may have consequences beyond food crop failures and food shortages. In 1998, with London, England, having the lowest rainfall record in 200 years, some tourist boats were unable to move about on the shallow River Thames. Water suppliers faced with low reservoir levels seriously considered building seaside desalination plants along the North Sea to make drinking water the same way it is done in the desert of Saudi Arabia. A drought can cut the amount of power generated by water-short hydroelectric dams, which in turn can lead to an energy crisis. Power rationing and a decrease in national production of commodities result. This is what occurred in Tanzania in December 1997 and Ghana in March 1998.

OTHER DROUGHT-RELATED PROBLEMS

Drought is the most serious physical hazard to agriculture in nearly every part of the world. Almost everywhere in the world soil moisture is deficient at some time of the year, depending primarily on the amount of rainfall or adequate water supply to the vegetation from irrigation systems. Even in humid climates such as the eastern United States, as in 1999, or western Europe droughts are frequent and severe. Since both rainfall and the demands for water by crops vary from one year to the next the magnitude of droughts also varies. For many years climatologists, meteorologists, and soil scientists have endeavored to forecast drought incidence and intensity with only limited success. Attempts at cloud seeding with dry ice and other means in order to generate precipitation were equally less than successful.

The drought that struck much of the United States in 1998 illustrated a number of ways people have become dependent on normal climate patterns. During the drought the Mississippi River's flow was so reduced that barge traffic had to be halted, hydroelectric output decreased, and wheat yields were reduced by more than half in the Great Plains. Food prices increased noticeably. Wildfires burned out of control in the Rocky Mountains, and residents of cities all across the nation had to reduce their water usage.

On June 26, 1999, with grazing lands parched after Arizona's third-driest winter in a century, the governor of that state declared a statewide drought emergency so that struggling ranchers could qualify for tax breaks. The drought followed a winter season of too little precipitation. Consequently, the lack of growth of grazing grass and dried up ponds caused many ranchers to sell their undersized livestock at market earlier than usual. In New Mexico the sparse rainfall meant many farmers had to plow under their very profitable chili crop, a source of 60 percent of the nation's base chili products.

Droughtlike conditions can cause water-use restrictions in both farm and nonfarm communities throughout the United States. Water use for irrigation can cost more, thus producing higher crop prices. But farmers, often in debt, may need to risk the avoidance of bankruptcy during a water crisis by starting their irrigation pipe pumps despite ordinances that aim to conserve limited water supplies obtained from rivers, reservoirs, and underground aquifers. This is what some farmers did in May 1996 during the extreme drought in the Southwest. Eight states were parched by rainfall amounts that were only 18 to 42 percent of normal. Agricultural losses can be so staggering during droughts that losses to farmers can soar to billions of dollars. When water emergencies are declared by government official, residents are barred from filling pools, sprinkling lawns, washing cars, or using water for nonessential reasons. Violators are usually fined.

Parts of the eastern seaboard; from Rhode Island to Maryland and West Virginia, endured the worst drought in decades—perhaps a century—during July and August 1999. Emergency measures on the restrictions of water use were put in force by the governors of several states in the region. New Jersey, the Garden State, with 9,400 farms, was designated a disaster area. That enabled distressed farmers there to apply for low-interest loans in order to keep their farms financially viable. The state received less than two inches of rain for well over a two-month period, compared to an average of more than eight inches for the same period in a normal year. The drought caused most kinds of crops to wither and lose vigor.

The dry spell in the Northeast affected businesses in different ways. Weather-dependent businesses, such as farming and landscaping concerns, experienced large losses. Paradoxically, the profits of others, like truckers with huge water tankers, well drilling companies, and boat repair shops located near recreational areas where the water levels hit thirty-year lows did well. It seems that numerous boaters damaged their crafts by running aground or striking rocks and sandbars in shallow waters.

The drought of 1999 brought some unusual concomitant effects. In New England, where the stunning, colorful fall foliage is a huge tourist attraction each year, the dry leaves fell early or turned a drab muddy yellow, brown, and orange. That discouraged many visitors accustomed to viewing the hillsides annual mosaics of bronze, scarlets, and purples. Trips to the region were canceled and business suffered considerably. The lack of rain also meant that those who enjoyed the outdoors had fewer mosquitoes to contend with, so bug spray manufacturers experienced lower sales. A most surprising result of the lengthy arid period was the postponement of a number of preseason football scrimmages at high school athletic fields. The grounds were judged to be much too hard—concretelike in some instances—and serious injuries to the players were feared.

RECORD-BREAKING DROUGHTS

Rainless days can last a long time. By September 4, 1997, Los Angeles broke a seventy-year record for days without any measurable rain, a dry spell of 198 days. San Bernardino County, California, went without rain from 1909 to 1912—1,001 days—followed by another period of consecutive rainless days. This was one of the longest periods of drought in the United States, causing extensive crop failures and damage to home landscapes and gardens. The world's largest recorded drought lasted 400 years. It took place in Claina, Chile, in the Atacoma Desert. From 1571 to 1970 not a single drop of rain fell during that time.

DROUGHT PREDICTION AS A GEOPOLITICAL TOOL

These days, American intelligence officials are examining new factors in the environment of troubled areas overseas that could result in the long-term causes of political unrest, wars, and disasters. They are becoming aware that outbreaks of clan warfare or ethnic violence may flare under the strains of hunger, drought, and a lack of arable land in conjunction with an explosion in population growth. The U.S. State Department, National Security Agency, and the Central Intelligence Agency are gathering "soft" intelligence data about water table levels as predictors of famine and the collapse of governments, as well as soil erosion, dwindling amounts of grazing land, and the expansion of deserts. They are also concerned with matters directly related to the global food situation, such as infant mortality and the rising tide of farm village people moving to the large cities of Africa, Asia, and Latin America, trends that are feared could produce many of the world crises of tomorrow.

DROUGHTS

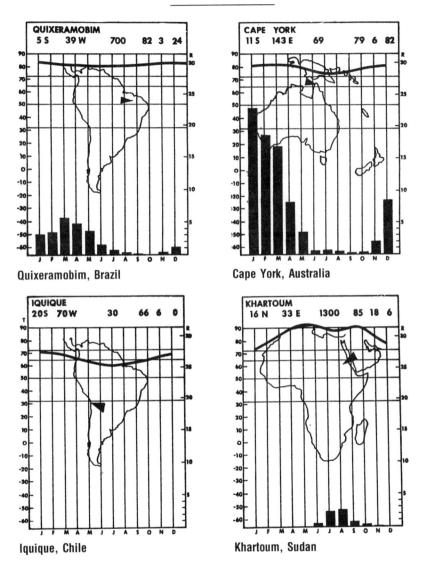

QUIXERAMOBIM
5 S   39 W   700   82 3 24

Quixeramobim, Brazil

CAPE YORK
11 S   143 E   69   79 6 82

Cape York, Australia

IQUIQUE
20S 70W   30   66 6 0

Iquique, Chile

KHARTOUM
16 N   33 E   1300   85 18 6

Khartoum, Sudan

Noting the rainfall curves (see key in the Introduction), the climographs indicate that these locations record very dry weather for long periods of time. The sparse amount of rainfall, accompanied by hot arid air over a tropical desert or savanna vegetation, presents many extreme hardships for those that live in that environment. The climate conditions people to be resourceful and adaptable if they are to survive the enervating elements.

SOURCES

Beveridge, Dirk. "Drought Parches Britain." *Sun-Sentinel* (Fort Lauderdale, Fla.). May 4, 1998.

"Brazil Calls in Shamans to Fight Amazon Fires." *New York Times*. March 30, 1998.

Crossette, Barbara. "Relief Teams Say North Korea Faces Vast Drought Emergency." *New York Times*. August 5, 1997.

"Drought Causing Famine." *New York Times*. December 16, 1997.

"Dry Spell Changing Outlook of Summer-oriented Business." *Home News Tribune* (East Brunswick, N.J.). August 7, 1999.

"Dry Weather in Plains States Bring Surge in Grain Prices." *New York Times*. May 11, 1997.

"Famine Threatening Children, Adults in North Korea." *New York Times*. March 19, 1997.

"Fear and Famine in North Korea." *New York Times*. August 8, 1997.

French, Howard W. "A Drought Halts Ghana On Its Road to Success." *New York Times*. March 16, 1998.

Glantz, Michael H. *Drought Follows the Plow*. Cambridge, England: Cambridge University Press, 1994.

Greenhouse, Steven. "The Greening of U.S. Diplomacy: Focus on Ecology." *New York Times*. October 9, 1995.

"Lack of Rain Cuts Off Power." *New York Times*. December 1, 1994.

"Los Angeles Sets Record for Rainless Days." *New York Times*. September 5, 1997.

Morinin, Farah A., et al. *Global Insights—People and Cultures*. New York: Glencoe, Macmillan/McGraw-Hill, 1994.

Natural Disasters of North America (supplement). *National Geographic*. July 1998.

Nordheimer, Jon. "The Fields, After a Month of Sun." *New York Times*. September 2, 1995.

Schemo, Diana Jean. "In Brazil, Despair Once Again Feeds on Drought." *New York Times*. May 28, 1998.

Schmieder, Allen A., ed. *A Dictionary of Basic Geography*. Boston: Allyn and Bacon, 1970.

Symons, Michael. "State Still Thirsting." *Home News Tribune* (East Brunswick, N.J.). July 19, 1997.

Verhovek, Sam Howe. "Parched Texas Rejoices In Hurricane's Aftermath." *New York Times*. September 4, 1996.

4

DUST STORMS

A dust storm occurs mainly in dry desert or semi-arid regions primarily in low- and midlatitudes. The hot moistureless air, filled with dust, is raised to great heights by turbulent winds causing the loose soil to be picked up and carried many miles away. The likelihood of this taking place increases as droughts linger in an area for long periods of time. A dust storm is often referred to as a sandstorm, especially in North Africa. Dust storms can be very unpleasant for a person's eyes, nostrils, or throat and might even cause temporary breathing problems. Some dust storms, driven by high winds and laden with gritlike materials, can cause physical damage to vehicles and buildings.

HISTORIC DUST STORMS

One dust storm that swept across Algeria in 1902 resulted in considerable particles of sand being carried about 1,100 miles to the British Isles. There is also evidence that specks of materials, wind-driven over the Sahara Desert by a dust storm, had been deposited beyond the Atlantic Ocean on the surface of numerous islands in the Caribbean Sea.

In the United States during the period 1935–1941, throughout much of the Great Plains region, a national calamity called the "Dust Bowl" took place. Farmers harvested good wheat crops during wet years. However, after a long-lasting drought, accompanied by stronger than normal winds, the topsoil began to blow away. Quantities of dusty soil landed as far away as the Atlantic Coast. During the Dust Bowl years of dust storm after dust storm, people living in the region were affected in a number of ways. Victims of the storms, sometimes called black blizzards, often found it impossible to see farther than a few feet in front of them. People resorted to wearing masks to protect their throats and lungs. The high winds eroded the loose soil, making any kind of planting useless. For years the prairie grazing lands were diminished in scope and visi-

During the period 1935–1941 a national disaster took place throughout the Great Plains and prairie lands of the western portion of the United States. Farm after farm, similar to the one pictured here, became unproductive waste lands when grazing acreage and field crops were blanketed by layers of wind-blown dust. Thousands of farmers during this Dust Bowl calamity suffered tremendous losses, causing many to cease operating and seek new opportunities in other places and occupations. (Courtesy of U.S. Department of Agriculture.)

bility. Dust clouds darkened the skies at midday, and the hot burning temperatures made the air seem furnacelike. Roads were obliterated. Farmsteads were partially buried under six-foot drifts of dust and sand. Machinery and vehicles were ruined. Trees planted as wind breaks became uprooted. Countless number of animals perished. People's lives were also lost.

During the 1930s tens of thousands of victims of the disaster were forced to relocate, many forming pathetic ranks of workers eager to do hard labor tasks in the orchards and vegetable fields of California. The story of the difficulties faced by many families in their migration became a classic part of American literature as written by John Steinbeck, in his book *The Grapes of Wrath*. The area devastated by the Dust Bowl covered some fifty million acres and included parts of Texas, New Mexico, Colorado, Kansas, and Oklahoma. Other parts of the Great Plains were affected as well, particularly the land eastward from the Rocky Mountains to an uncertain line where rainfall averaged only twenty inches per year.

RECENT DUST STORMS

The state of Arizona is still plagued by dust storms, especially during early springtime. In April 1997 swirling dust along a highway in the southeastern part of the state caused a twenty-four-car pile-up, which resulted in the deaths of ten people. The police reported the visibility to be less than a car length during the storm. The stretch of highway where the accident occurred is considered one of three dust traps in Arizona. In 1989, another blinding dust storm caused a thirty-one-car accident that killed forty and injured over forty. Now large yellow signs warn motorists of blowing dust along the roads where dust storms have been a problem. A resident of the area who experienced the storm, likened the effect "like being sandblasted." She reported that "when the sand hits you, it burns you, it gets in your eyes. You don't want to be out in it." Arizona averages about thirty dust storm–related accidents a year.

DUST STORMS

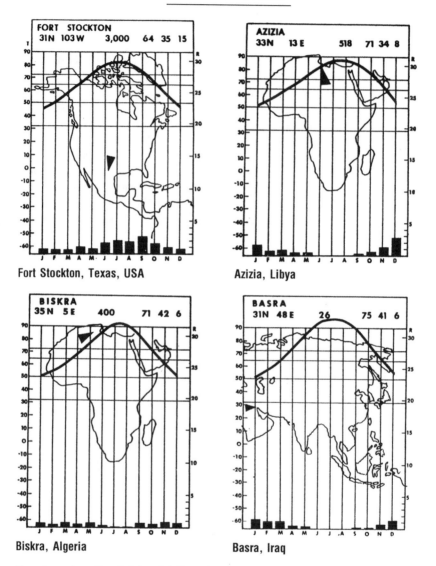

FORT STOCKTON
31N 103W 3,000 64 35 15

Fort Stockton, Texas, USA

AZIZIA
33N 13E 518 71 34 8

Azizia, Libya

BISKRA
35N 5E 400 71 42 6

Biskra, Algeria

BASRA
31N 48E 26 75 41 6

Basra, Iraq

By observing the total rainfall and average temperatures (see key in the Introduction), you can determine that these weather stations provide typical temperature and rainfall data normally found in tropical deserts and tropical steppe climates. Incessant high winds can readily blow away loose grains of sand and dried-out topsoil, causing large dark clouds of earth particles for miles around. This phenomenon results in crop-growing failures and makes human habitation most discomforting, prompting many farmers to leave that area and never return.

SOURCES

Borchert, J. R. "The Dust Bowl in the 1970s." *Annual Association of American Geographers* 61, 1971.

Carrol, Ginny, and Peter Annin. "Another Dust Bowl?" *Newsweek*. July 1, 1996.

Hughes, Patrick. "Dust Bowl Days." *Weatherwise*. June/July 1995.

National Disasters of North America (supplement). *National Geographic*. July 1998.

Oliver, J. E. *Climate and Man's Environment*. New York: John Wiley, 1973.

Oliver, John E., and Rhodes W. Fairbridge, eds. "Dust Storms." *The Encyclopedia of Climatology*. New York: Van Nostrand Rheinhold, 1987.

Parkinson, G. R. "Dust Storms Over the Great Plains." *Bulletin of the American Meteorological Society*. May 1956.

Schneider, Stephen H., ed. *Encyclopedia of Climate and Weather*. New York: Oxford University Press, 1996.

"10 Killed in 24-Car Pileup in Dust Storm." *New York Times*. April 7, 1995.

Warm, G. F. "Some Dust Storm Conditions on the Southern High Plains." *Bulletin of the American Meteorological Society*. June 1952.

White, C. Langdon, and George T. Renner. *Human Geography*. New York: Appleton Century-Crofts, 1936.

White, C. Langdon, George T. Renner, and Henry Warman. *Geography: Factors and Concepts*. New York: Appleton-Century-Crofts, 1968.

People residing in low-lying flood-prone areas most often have little immunity against the devastation that could be brought on by periodic flooding. Here a mother and young son in March 1990 ponder the damage to their Elba, Alabama, home as they patiently wait and watch the unwelcomed waters recede. (Courtesy of American Red Cross.)

5

FLOODS

Floods take place when normally dry land is covered, or inundated, by water. People have suffered from destructive floods for over 4,000 years of recorded history. Since the beginning of time floods have been the most devastating of natural forces known to man. Floods occur when waters rise above a level of containment in a natural stream or man-made trough. Once the overflow spills over its bank some degree of damage usually results. The damage may be confined to the adjacent local area or it may extend miles from the immediate flood site.

THEORIES AND CAUSES

Besides the weather factor, the location of a place contributes to the possibility of flooding. Low-lying areas, such as flood plains, river basins, and coastal plains are often vulnerable to widespread flooding. Various kinds of riverbeds, such as arroyos and wadis, coastal regions, and valleys subject to rain or melted snow run-off are especially prone to flash flooding. Floods are most commonly attributed to ongoing rains, local thunderstorms, hurricanes, tidal waves, typhoons, cyclonic disturbances, and melted snow and ice. Tremendous pressure of swollen rivers can cause water bodies to spill over levees and embankments, causing staggering loss of life and extensive damage to property in the afflicted area. When sea walls, flood gates, dikes, or other kinds of man-made retaining structures designed for controlling the volume of water in reservoirs, dams, and canals break or leak, a major flooding calamity can result. Ninety percent of all disasters in the United States are flood-related. A third of all flood losses in America take place outside flood hazard areas.

HISTORIC FLOODS

The first meterologically recorded flood took place in Holland in 1228 with the loss of 100,000 lives. In the Netherlands, there is a long and frustrating history of battling ruinous floods. In 1953 a North Sea dike burst killing more than 1,800 Dutch citizens. China is another country with a disastrous past in regard to floods. The greatest flood ever to occur there was in August 1931, when the Huang Ho River overflowed, killing 3,700,000 Chinese. The total casualties would be comparable today to the loss of half of New York City's population. Prior to that horrendous catastrophe, the same river claimed the lives of 900,000 from an 1887 flood. Over the years, hundreds of thousands of Chinese peasants have perished during periodic flooding of the countryside. India, Pakistan, and Bangladesh have experienced notable floods and tidal waves that have caused countless deaths. Tidal waves are unusually great destructive waves sent ashore by an earthquake or a very strong wind.

The United States has hardly been spared the rage and fury of flood episodes. Floods have "visited" every state. On May 31, 1889, the most destructive flood in American history took the lives of 2,209 residents of the Johnstown, Pennsylvania, area. Another flood there in July 1977 cost the area seventy-six lives and did $330 million in damages. From September 6–20 1928, it is estimated that more than 1,836 people in Florida died when Lake Okeechobee overflowed its levees and flooded a wide region during a hurricane. On March 21, 1913, heavy rains caused widespread flooding of the Ohio River Basin, resulting in the deaths of 467 people.

In 1993 a "Great Flood" along the Mississippi River spread over eight million acres of land and caused more than $12 billion in damages, along with fifty deaths and left 70,000 people homeless. A tremendous deluge of rainfall in South Dakota killed 242 people in a short period of time on June 9, 1977. An unprecedented downpour once caused rivers in central Arizona to rise as much as ten feet per hour, sweeping cars and buildings thirty to forty miles downstream. In Kansas City, Missouri, twelve inches of rain fell in twenty-four hours bringing five feet of water rushing through the streets.

In March 1997 a flood along the Ohio River Valley caused thirty-five people to lose their lives. The rising Red River drove residents of Grand Forks, North Dakota, from their homes, many of which were destroyed, during a period starting on April 19, 1997.

DEALING WITH FLOODS

Floods, like other stressful natural disasters, can bring out the best and worst in people. News reports about flood-stricken towns and cities reveal that some individuals selfishly disobey the orders of authorities to evacuate their homes and businesses, refuse to assist others, especially the elderly, in life or death situations, and even resort to looting flood victims' homes and property.

On the other hand, the many good deeds and heroics of civilian volunteers and concerned citizens are provided as splendid examples of positive human relationships. People, even small children, voluntarily fill sandbag after sandbag in their attempt to build dikes meant to hold back rising floodwaters. Many assist others in removing important household furnishings from neighbors' homes in advance of the encroaching waters. All kinds of boats, rubber rafts, and suitable vehicles are offered to evacuate stranded people from flooded homes and rooftops. They are also used to alert residents of flood-threatened neighborhoods of impending danger and to search for victims.

People from unaffected areas, some from distant locations, have volunteered their services in many helpful ways. News reports consistently show cases where flood victims are invited to stay in homes of strangers with gracious hospitality extended. Unknowns have deposited food and safe water on the porches of homes where residents have returned to begin to reestablish their lives. Volunteers have worked long hours in emergency shelters offering every kind of comfort for those in need, be it bedding, clothing, nursing, food serving, counseling, or recreational provisions.

Once the flood is over many other kinds of serious problems persist for flood victims before normal conditions are resumed. Entire communities may have to examine and repair their infrastructure, including roads, schools, power plants, transportation networks, and communication systems. Health officials may need to check local sewage disposal facilities and water supplies. They will also have to be diligent in monitoring the possible danger of epidemics being spread. Building inspectors will need to evaluate damaged homes before permission is granted for reoccupancy. Insurance claims have to be settled quickly for victims short of funds. Governmental aid and bank loans need to be offered at low costs to those who choose to restore their existing homes.

In the aftermath of a devastating flood, when sanitary controls are depleted or unavailable, a rapid epidemic of disease can result. This can happen when medicine is in short supply or when inoculation of the flood victims is inadequate. Uncollected garbage accumulates, pools of stagnant bacteria-ridden water form, drinking water becomes contaminated, and mosquitoes and rodents carry germs to a weakened population. Therefore, flood effects can linger for weeks following the disaster.

PRECAUTIONARY MEASURES

In some flood-prone areas the government is now offering attractive financial incentives to residents to abandon their previous home sites and rebuild their homes at higher levels. After the devastating floods of 1993, the entire town of Valmeyer (900 residents) in southern Illinois moved their dwellings one and a half miles to the east, on a bluff over-looking the Mississippi River. Thousands of others along riverbanks, tired of contending with frequent flooding, have taken advantage of this program.

In China, the world's most populated country, floods have proven to be a major problem, particularly along the Yellow River, where in the past millions have been killed. That is why the river is still referred to as "China's Sorrow." Now in a colossal plan to manage the flow and uncertainty of the Yellow River, a controversial $3 billion dam is being built at Xiaolangdi to be completed in 2002. Flood control is a major goal of the project, but its construction will force the relocation of 400,000 peasants presently working the land there. Still, the Chinese continue to fight to hold back their rivers from flooding. In July and August 1998 the 3,900 mile Yangtze River, swollen to record high levels, damaged dikes along storm-ridden areas where more than 3,000 are said to have died. More than fourteen million people have had to abandon their homes in the summer flood season. Flood threats in China have affected 240 million people, a fifth of the population.

PREPAREDNESS AND PREVENTION OF DISASTERS

A five-year study costing $750,000, commissioned by the National Science Foundation and various federal agencies of the United States, found little long-term planning in most areas of the country having the highest risks for natural disasters. Disasters appear to be striking more populated areas than ever before in the nation's history. Governmental reforms and legislation efforts are beginning to focus on preparedness and prevention, rather than costly emergency aid, in their introduction of new policies.

The United States has spent more than $500 billion on disasters from 1979 to 1999. Florida, California, and Texas had the brunt of the natural calamities. Now those states and others are beginning to make their building codes more effective. Federal flood insurance programs are now offering below-normal premiums to property owners in flood-prone regions. The government is also buying properties in vulnerable areas where repeat disasters tend to strike, forcing homeowners to build out of harm's way.

FLASH FLOODING

A surprise deluge of torrential rains fell throughout the New York metropolitan area during the morning commute period on August 26, 1999. For two hours as much as four inches of rain flooded the region crippling for much of the day highways, railways, subways, and streets, causing all kinds of traffic delays, disruptions, and bottlenecks. It was the worst flash flooding in years. It was estimated that four million people were either stranded or unable to reach their jobs, many reporting to their workplaces hours late.

The rains submerged highways and train tracks. In low-lying regions "ponds" of water were nearly five feet deep. Dozens of commuters were trapped in their cars and had to be rescued. Hundreds of thousands were trapped on trains, buses, and subway platforms. Hundreds of storm sewers were backed up. Along many roadways geysers popped off manhole covers. For the very determined they merely rolled up their trousers or slack bottoms, removed their shoes, and waded on to work, arriving with a litany of adventures to share.

FLOOD IN LITERATURE AND FOLKLORE

Stories of great floods in ancient times are legendary. The Bible story of Noah and the ark has been told from generation to generation. It is an account of how Noah was able to preserve the species when waters rose for 150 days to cover the earth. Every living thing was destroyed except for those sheltered in the ark. The story is most often revealed as a moral lesson. Various ancient cultures, going back to the Assyrians, Babylonians, and Persians, have adapted different versions to the epic. Some scholars cite archaeological facts to conclude that there were many floods during early times, particularly in the Mesopotamian lands of the Tigres and Euphrates Rivers. Legendary narratives refer to the cataclysms of that era.

There are many descriptive accounts of floods in Greek mythology and the folklore of Europe and southern Asia. The story of the flood is mentioned in Greek literature as early as the fifth century B.C. The popular mythology of Zeus submerging Greece in torrential rains and how King Deucalion and his wife Pyrrha take refuge in an ark, who upon being saved, choose to increase the human race as a wish fulfilled, has been retold centuries on end.

Over the years the Dutch people have been highly impacted by the legend and folklore of their flooded lands, often brought on by collapsed dikes unable to hold back the force of menacing sea waters. There is the classic character Pieter featured in Mary Mapes Dodge's book *Hans Brinker* who was credited with an event that never really happened. The young lad supposedly saved his village from a flood by putting his finger in a hole in a dike, holding it there all night until repairs could be made. So many tourists to Holland inquire about the legendary hero that a statue in his honor was erected in the small village of Spaarndam, a few miles north of Haarlem.

FLOODS

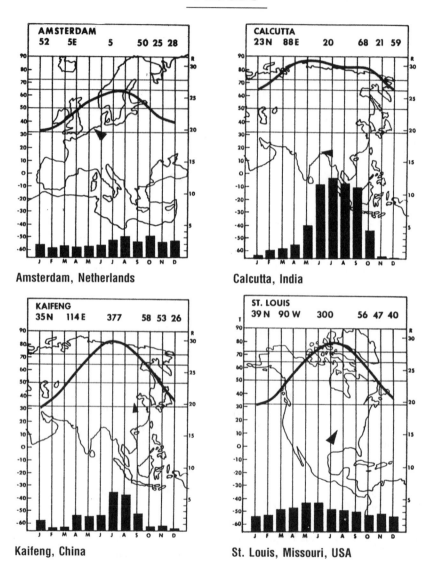

AMSTERDAM				
52	5E	5	50 25 28	

Amsterdam, Netherlands

CALCUTTA				
23N	88E	20	68 21 59	

Calcutta, India

KAIFENG				
35N	114E	377	58 53 26	

Kaifeng, China

ST. LOUIS				
39N	90W	300	56 47 40	

St. Louis, Missouri, USA

Each of these cities has a long history of flooding, which is indicated by the total rainfall (see key in the Introduction). Previous flood disasters have taken hundreds of thousands of lives and caused untold accounts of victims being displaced in these regions. Property damages have been calculated in the billions of dollars. Despite repetitive flooding of these areas there tends to be a human inclination for many to return to their homes and businesses and to once again endeavor to rebuild their lives at their previous flood sites.

SOURCES

Costa, J. E. "A Comparison of the Largest Rainfall—Runoff Floods in the United States with Those of the People's Republic of China and the World." *Journal of Hydrology* 96 (1987).

Cowell, Alan. "With Nearly 100 Dead, Floods Keep Raging in Central Europe." *New York Times.* July 21, 1997.

"European Floods Inundate Netherlands," *Sun-Sentinel* (Ft. Lauderdale, Fla.). February 1, 1995.

"Flash Floods and Floods—the Awesome Power," A Preparedness Guide, National Oceanic and Atmospheric Administration, National Weather Service, July 1992.

"Floodwaters Wreak Havoc in Spain," *Home News and Tribune* (East Brunswick, N.J.). November 7, 1997.

Griffith, Martin. "50,000 Evacuated in California Cities. At Least 16 Deaths Blamed on Weather." *Sun-Sentinel* (Ft. Lauderdale, Fla.). January 3, 1997.

Janofsky, Michael. "Residents of Flood Plains Balk at Moving." *New York Times.* March 3, 1997.

McClure, Robert. "Indian Canyon Village Ravaged by Freak Floodwaters in August." *Sun-Sentinel* (Ft. Lauderdale, Fla.). December 7, 1997.

McFadden, Robert D. "Surprise Deluge Cripples Morning Rush in New York." *New York Times.* August 27, 1999.

Meredith, Robyn. "States Hit by Flooding Focus on the Threat of Disease." *New York Times.* March 11, 1997.

Natural Disasters of North America (supplement). *National Geographic.* July 1998.

Oliver, John E., and Rhodes W. Fairbridge, eds. *The Encyclopedia of Climatology.* New York: Van Nostrand Rheinhold, 1987.

Page, Trevor. "Flood Waters Come and Go. Misery Endures." *New York Times.* May 26, 1996.

Petterson, Roger. "A Raging River—Swollen Ohio has Thousands Evacuating." Sun-Sentinel (Ft. Lauderdale, Fla.). March 5, 1997.

Sack, Kevin. "Storm's Rains Bring Flooding in Two States." *New York Times.* September 8, 1996.

Schneider, Stephen H., ed. *Encyclopedia of Climate and Weather.* New York: Oxford University Press, 1996.

Terry, Don. "For a Flood-Stricken Town, Hope Is a Next-door Neighbor to Despair." *New York Times.* May 14, 1997.

———. "Wading Through Mud and Sadness in Search of Hope." Sun-Sentinel (Ft. Lauderdale, Fla.). March 9, 1997.

"Thousands Flee Flooding: Governors Declare Emergency." *Sun-Sentinel* (Ft. Lauderdale, Fla.). January 4, 1997.

Tyler, Patrick E. "China's Fickle Rivers, Needy Industry Brings a Water Crisis." *New York Times.* May 23, 1996.

U.S. Department of Commerce. *Flash Floods and Floods, The Awesome Power!* Washington, D.C.: NOAA, 1992.

"Volunteers Pouring In to Assist Communities Ravaged by Floods." *Los Angeles Times.* March 11, 1997.

Walsh, Edward. "With Flood Waters at Their Door. A Few Stubborn Souls Ride It Out." *Washington Post.* March 8, 1997.

———. "Unpredictable Nature Inundates Many Towns: Flooding Hits Some Hard, Spares Others." *Washington Post.* March 7, 1997.

"Worst Flood in Two Centuries—Hundreds Flee to High Ground as Dike Succumbs." *Home News and Tribune* (East Brunswick, N.J.). July 26, 1997.

6

FOG

Fog is essentially a cloud at the earth's surface, or sometimes just a few feet above. It consists of numerous small droplets of water, not readily or clearly perceived by the naked eye. Fog is the one element of nature that still baffles man. When it shrouds an area, it envelops everything in a kind of gray vapor, dimming or blurring the vision. Fog can be dense when objects are obscured at a thousand feet or less. Light fog has increased, but still hindered, visibility. Fog begins to develop when the air is cooled below its dew, or saturation, point, which is a temperature below the capacity of the air to hold water. Fog can form on land or sea.

During clear nights heat absorbed by the ground during the day escapes rapidly or radiates into the atmosphere, usually late in the afternoon. This chills the air near the ground to dew point and results in a still fog called *radiation fog*. This type of fog develops when there is moisture in the air and wind speeds are calm or light. Another kind of fog is known as *advection fog*. This classification forms when warm air blows across a cold land or water surface. The air is chilled and the moisture condenses as fog. Radiation fog is often found in valleys where winter moisture gets trapped within the valley walls. Advection fog tends to occur when moist air passes over a colder surface, such as a cold ocean current, or over an area covered by ice or snow. It is possible for a sky to be observed during a ground fog settling in a low lying area or mountain hollow.

EFFECTS OF FOG

Fog can cause considerable anxiety and disruption for people engaged in normal living activities. They have lost time caused by traffic delays. Trains and other forms of public transportation have had to be canceled or rescheduled. Highways have either been closed or had their speed limits reduced during thick fog conditions. Sometimes airports are shut.

Fog reduces visibility sometimes for days. Usually all means of transportation are affected to some degree. Here a fleet of commercial fishing boats, unable to rely on visual navigation, is forced to remain at dockside to await a clearing of the weather. This results in loss of income for the fishermen. (Courtesy of National Oceanic and Atmospheric Administration.)

Aircraft are at times "socked-in" by fog, and planes are frequently ordered to land at airports having more favorable visibility reports. Schools have been closed, or students let out early during fog development. All kinds of performances, special events, and community meetings have been postponed when people are unable to attend the affairs because of restricted visibility on the roads. Fog always brings on the threat of multivehicle accidents. A record pile-up occurred near Padua, Italy, on February 11, 1998, when 250 cars and trucks crashed on a fog-ridden highway killing four motorists and seriously injuring dozens.

Prior to the advent of radar and recent state-of-the-art navigational aids being used on sea-going vessels, the toll from boat accidents had been extremely high due to foggy conditions. Tens of thousands of sailors and passengers have drowned due to ships ramming other vessels, striking rocks, running aground, or landing on a reef during dense fog. Over the years captains and quartermaster pilots on a ship's bridge have relied on fog horns and ship bells to alert other vessels of their proximity during fog. Their warning sounds have not always been heard or heeded.

Many tragic unusual accidents on water have resulted during fog when the sight of a ship's pilot or quartermaster has been obscured. Subsequent investigations of the accidents also attribute losses at sea during fog to carelessness, faulty navigation information, failure to avoid foggy locations, and excessive speed in precarious fog alert zones reported by radio transmission from other ships. In addition, carelessness and disregard of radar surveillance, along with ship personnel not sounding or heeding warning signals in fog conditions, is blamed for fog-related accidents.

THE FOGGIEST PLACES

Because of specific site and situations, some people live with more foggy days than others. Generally the drier the location, the less fog is anticipated. Cape Disappointment, Washington, is the foggiest place on the West Coast of the United States. There people are confronted with an average of 2,552 hours of fog per year. That means that residents there are in fog about 30 percent of the time. And those living near Moose Peak Lighthouse on Mistake Island, Maine, face 1,580 hours per year. It is the foggiest place on the East Coast of the United States.

In the United States the highest average number of days each year with heavy fog is recorded along the Pacific Coasts of the states of Washington and Oregon. Those locations, as well as the Maine coast and the mountains of Tennessee, Virginia, West Virginia, New Hampshire, and Vermont, all have more than eighty days of fog each year. Generally the least number of foggy days (under ten) are found in the intermontane west of America.

Sea fog is a common occurrence along ocean shores wherever warm currents merge with cold currents, such as is the case in the northern Pacific Ocean region off the coasts of Alaska and Canada, and the New England and Newfoundland shores of the northern Atlantic Ocean. Sea fog conditions are also found in the high latitudes of the Atlantic, as well as the seas northeast of Japan.

In warfare, the forming and lifting of fog have led to military successes and failures. Warships have hidden their positions and visibility behind protective curtains of fog. With good forecasting they were able to launch a timely attack upon the exposed enemy fleet once the fog lifts. Ground troops use the cover of fog to change their positions, bring supplies to the line, strike out with an offensive, or retreat if need be, knowing the fog will help shield them from opposing incoming fire. And airmen, in their flight plans for bombing raids on enemy targets, prefer a clear sky for dropping their bombs. Once their mission is completed, they may, if pursued by enemy squadrons, seek out thick clouds or dense fog in order to out-maneuver their hostile opponents.

FOG IN LITERATURE AND FOLKLORE

Anyone who has ever experienced foggy conditions will remember for a long time his or her reaction and the feeling of restricted sight and mobility. The atmosphere was both misty and mystifying, indeed scary.

The word fog has developed into a descriptive reference in our English language usage. We refer to such terms as "in a fog," and "haven't the foggiest idea," or a "pea soup fog," along with "fogging-up" (windows) and "fog bound," now common phrases to underscore such foglike traits as vagueness, muddled, dim, or unclear. Descriptions of fog have served to provide background settings for mystery novels. Fog scenes are often used when authors wish to establish a frightful tone in a story. Charles Dickens in his *Christmas Carol* describes a fog in London, a city once renowned for fog so thick that you could get lost in it while crossing the street.

FOG

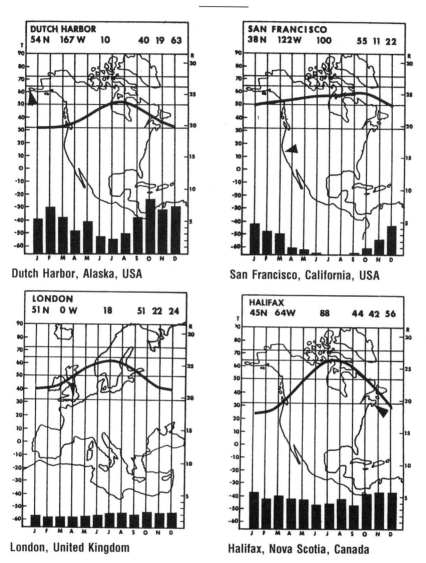

Dutch Harbor, Alaska, USA

San Francisco, California, USA

London, United Kingdom

Halifax, Nova Scotia, Canada

When fog occurs, as it frequently does in places such as those depicted, due to their locations (see key in the Introduction), where moisture-laden air flows in from the nearby sea, disruptions and slow-downs result in normal living activities. Limitations in visibility can readily alter or restrict travel schedules, parcel and cargo deliveries, work attendance, business meetings, social engagements, and cause serious accidents for all modes of transportation.

SOURCES

Fisher, Robert Moore. *How to Know and Predict the Weather*. New York: New
    American Library, 1953.

Flohn, Herman. *Climate and Weather*. New York: McGraw-Hill, 1969.

"Four dead in 250-car pileup." *Sun-Sentinel* (Ft. Lauderdale, Fla.). February 11,
    1998.

Hess, W. N., ed. *Weather and Climate Modification*. New York: Wiley, 1974.

McIlneen, R. *Basic Meteorology*. New York: Van Nostrand Reinhold, 1986.

Moran, Joseph M., and Lewis W. Morgan. *Essentials of Weather*. Englewood Cliffs,
    N.J.: Prentice-Hall, 1995.

National Disasters of North America (supplement). *National Geographic Society*. July
    1998.

Santaniello, Neil. "Fog Shuts Down PBIA for 2 ½ Hours; More Likely Today."
    Sun-Sentinel (Ft. Lauderdale, Fla.). February 23, 1996.

Schneider, Stephen H., ed. *Encyclopedia of Climate and Weather*. New York: Oxford
    University Press, 1996.

Schneider, Stephen H., and Randi Louder. *The Coevolution of Climate and Life*. San
    Francisco: Sierra Club Books, 1984.

Silverman, B. A., and A. I. Weinstein. *Weather and Climate Modification*. New York:
    Wiley, 1974.

7

HAIL

Hail is solid precipitation that falls as lumps or balls of ice. The individual units that fall to the ground must be at least five millimeters (0.2 inches) in its longest dimension to qualify as hail. The dynamics of hail formation starts with strong vertical air currents moving rapidly upward in warm weather. The overheated air carries large raindrops into cold air aloft that freezes them into ice crystals. As the hailstones become heavy they drop only to again be lifted by high-speed updrafts, estimated to be close to 35 miles per hour. During the process of being carried up and down many times between the cold and warm layers of air, concentric rings of ice and snow build over one another. The coated rings of ice and snow in a hailstone resemble a dissected onion. Once the hailstone becomes extra large and overly heavy it is unable to stay up in the air, and it then falls to the ground. The speed of a falling jumbo-sized hailstone is estimated to be forty-five meters per second, or 105 miles per hour.

HAILSTONE SIZE

Sometimes sport fishermen tell tall tales about the size of a fish they once caught. To some degree "hail sighters" also tend to exaggerate about the size or weight of a hailstone they once came across. Allegedly the largest single stone ever measured in the United States fell at Potter, Nebraska, on July 6, 1928. It weighed 1 1/2 pounds and was seventeen inches in circumference and almost 5 1/2 inches in diameter. On September 1970, in Coffeeville, Kansas, a hailstone having equally formidable measurements motivated its excited townspeople to enter a local hailstone finding competition. But, the prize-winning stone supposedly also measured five inches in diameter.

A hailstone similar to the one shown can be quite large and heavy. Allegedly the largest single stone ever measured in the United States fell at Potter, Nebraska, on July 6, 1928. It weighed $1\frac{1}{2}$ pounds, was seventeen inches in circumference and almost $5\frac{1}{2}$ inches in diameter. (Courtesy of National Oceanic and Atmospheric Association.)

EFFECTS OF HAIL

There is a kind of natural curiosity for people of all ages to want to rush outside in the immediate aftermath of a thunderstorm to try to find hail or to explore damages. A hailstorm, more than other kinds of atmospheric disturbances, is unique and fun to investigate. Hailstones of all sizes and shapes are strewn about, not unlike what could have been a glacial till deposit eons of time past. Yet, moments prior to a thunderous hail dispensing cloudburst no traces of snowy ice lumps are visible on the landscape. To be in an automobile during a hailstorm is an unforgettable experience. Watching the hail ricochet off the windshield, timed with deafening sounds similar to those blaring in a shooting gallery, can instantly grab your attention. Most thunderstorms only produce lightning, thunder, and rain. Approximately one in four hundred is so violent that it generates hail as well.

Hailstorms have the capacity to destroy farm crops and may even kill large animals. Hail at least the size of marbles from a storm ripping into Medford, Oregon, on September 4, 1997, destroyed 20 percent of the pear crop in that area's orchards. With the potential for hailstorm damage becoming so widespread and great, it is not surprising that farmers and rural people are increasingly purchasing hailstorm insurance coverage.

In the 1980s severe hailstorms inflicted substantial damage, injury, and death in Brazil, Germany, and China, as well as in Colorado in the United States. In some vulnerable areas nets are now being used to protect greenhouses and valuable crops. On July 11, 1990, a hailstorm in the Colorado Rockies caused $625 million in damage. Hail can clog sewers, make driving a car treacherous, and cut down visibility. It is obvious that no "hail"-like greeting was in order for the pellets of ice and hard snow that fell to the ground on that day in Colorado.

HISTORIC HAILSTORMS

On April 30, 1988, the world's most deadly hailstorm killed 230 people in Moradabad, India. In contrast, the greatest hailstorm casualties for a single day took the lives of two people near Lubbock, Texas, on May 13, 1930. On June 9, 1959, hail bombarded Seldon, Kansas, for a record eighty-five minutes, leaving hail eighteen inches deep. Hailstones ranging in size from baseballs to grapefruits accompanied by a 100-mile-per-hour windstorm once caused $200 million in damages in Dallas, Texas. On March 7, 1998, near Rogers, Louisiana, local residents reported that chunks of hail smashed windows and dented cars. One man there displayed a six-inch hole in his porch, punched in by a large-sized hailstone, which to him seemed like a grand finale to ten or fifteen minutes of a storm of nonstop thunder and lightning.

In the past the fear of a harmful hailstone prompted ignorant farmers to try to thwart the oncoming storm, hopefully driving it away with prayer and sacrificial offerings. It was commonly believed that the hailstorm was sent by the deities bent on punishing sinners and those deemed defiant. Later guns and gunpowder weapons were used in an attempt to destroy the hail while it was aloft.

PREDICTING HAIL

Forecasting hailstorms is extremely difficult. There are several techniques for detecting hail by radar. Hail forecasters rely on the same methods for predicting intense storms, however, they take into consideration the potential buoyancy of the raindrop or hail, the height of freezing temperatures, and the amount of moisture in the air aloft as determined by a wet-bulb hygrometer reading. Based on historical weather analogies it is anticipated that some areas have a higher prospective frequency of hail occurrence. The areas that generally receive more frequent hailstorms are situated on high terrain on the leeward sides of mountain ranges, such as the Rocky Mountains, the Alps, and Caucasus range. The highlands of Kenya is another site of considerable hailstorms. The leeward sides generally have heavy descending air that tends to pick up moisture as it drifts downward, thus giving and impetus to hail formation. The wheat and corn crops of the Great Plains are particularly vulnerable to hail in the United States. Grapes, tobacco, citrus fruits, tea, and vegetable crops in other parts of the world have been greatly damaged by hail.

HAIL

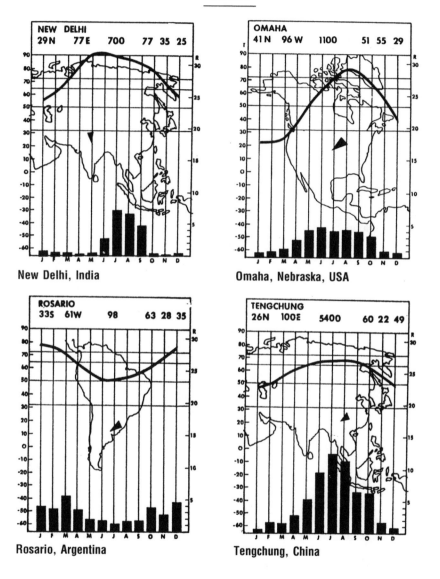

New Delhi, India

Omaha, Nebraska, USA

Rosario, Argentina

Tengchung, China

Hail falling is generally deemed to be a fairly rare weather occurrence, but some areas, such as those represented in the graphs, have a higher prospective frequency of occurrence due to fluctuations in temperature (see key in the Introduction), arousing the curiosity of those unfamiliar with hail to want to inspect the hailstones firsthand. In reality a hailstorm can cause considerable damage to crops, buildings, automobiles, and even take the lives of people. Anyone ever "pelted by pellets" of hail will probably recall the fury of their counter for a very long time.

SOURCES

Byers, H. R. *General Meteorology* (4th ed.). New York: McGraw-Hill, 1974.

Cotton, William R., and Richard A. Anthes. *Storm and Cloud Dynamics*. San Diego: Academic Press, 1989.

Fisher, Robert Moore. *How to Know and Predict the Weather*. New York: New American Library, 1953.

"Hail Puts Dent in Travel." *Sun-Sentinel* (Ft. Lauderdale, Fla.). March 8, 1998.

Houze, Robert A. *Cloud Dynamics*. San Diego: Academic Press, 1993.

Hughes, Patrick, and Richard Wood. "Hail: The White Plague." *Weatherwise*. April/May 1993.

"Large Hail Damages Pear Crop in Oregon." *New York Times*. September 5, 1997.

Ludlum, David M. *The Weather Factor*. Boston: Houghton Mifflin, 1984.

McIlneen, R. *Basic Meteorology*. New York: Van Nostrand Reinhold, 1986.

Natural Disasters of North America (supplement). *National Geographic*. July, 1998.

Oliver, John E., and Rhodes W. Fairbridge, eds. *The Encyclopedia of Climatology*. New York: Van Nostrand Rheinhold, 1987.

Wood, Richard A. *The Weather Almanac* (7th ed.). Detroit: Gale Research, 1996.

Trewartha, Glenn T., and Lyle H. Horn. *An Introduction to Climate* (5th ed.). New York: McGraw-Hill, 1980.

Schaefer, U. J., and J. Day. *A Field Guide to the Atmosphere*. Boston: Houghton Mifflin, 1981.

Schneider, Stephen H., ed. *Encyclopedia of Climate and Weather*. New York: Oxford University Press, 1996.

8

HUMIDITY

Weather reporters constantly refer to the "relative humidity." This is a reference term that cites the amount of water vapor in the air compared to the amount the air could hold at a given temperature and location. Warm air has a greater capacity to hold moisture than cold air. It is measured by a weather instrument called a hygrometer. The degree of wetness or dryness is expressed by a percentage figure, which indicates the amount of moisture required in the air for saturation at a specific temperature. Thus, if the air has only half the amount of vapor the relative humidity is 50 percent. Complete saturation of the air is 100 percent, which is a rarity unless it is raining.

EFFECTS OF HUMIDITY

"It's not the heat, it's the humidity." This is a true statement made by most people when their bodies tend to feel the damp heat in the air, which can make them listless and uncomfortable. When the air, laden with high humidity, becomes sticky the skin's pores open and the body tends to perspire more freely. Insects are attracted to body sweat and this becomes a major concern when their biting and feeding enable them to transmit diseases.

People who must cope with excessive heat and humidity can easily become irritable, short-tempered, and lose their appetites. Workers and others required to spend many hours out-of-doors during the hottest months of July and August (sometimes referred to the "dog days of summer") can lose a considerable amount of weight by sweating. On the other hand, even higher temperatures with much less moisture content in the air afford greater ease in functioning and a greater tolerance for the drier heat. Yet, people prefer one or the other weather situation depending upon what their own body feels—their physical sensation—frequently referred to as the sensible temperature.

Hot temperatures can send people swarming to the beaches, as pictured here. A hot and humid air mass blanketed the state of New Jersey in July 1999, and people sought sea breezes and the ocean for swimming at Margate Beach. (*Credit:* AP/Wide World Photos. Reprinted with permission.)

Humans, by the way their bodies react to the relative humidity index, can offer some insight into the amount of water vapor saturating the air at a given time. Older individuals may complain that their arthritis or rheumatism is "acting up." Their aches and pains become more pronounced. Human hair can reveal much about the amount of moisture in the atmosphere. Hair, like rope, swells in moist conditions and shrinks when the air is dry. As the relative humidity increases from next to nothing to 100 percent, the human hair expands 2 and one-half percent in length. That is why as early as 1783 inventors used human hair in making hygrometers. Barbers and wig makers, especially those who serve women, are well aware of this phenomena. Some people discover that their hair has a tendency to curl during days of high humidity and conversely tends to be straight during periods of dry weather.

A woman living in dry Arizona for many years may find her skin to be drier and more wrinkled than her twin sister's who lived most of her life in the more humid eastern part of the United States.

Around the home it is apparent that when the relative humidity is high, windows stick in the frames, clothes dry slowly, and salt cakes. Relying on fans and air conditioning is almost mandatory for comfort. When the air at home is dry, evaporation is rapid, soil dries out, and wet garments dry quickly.

PRECAUTIONARY MEASURES

During a scorching three-day road-buckling heat wave that started July 4, 1999, the weather reporting media throughout the northeastern section of the United States, where a number of temperature-humidity records were broken, broadcast a number of ways to cope with the un-relenting hot weather of that period. They advised people to (1) try to stay indoors in an air conditioned home or apartment, (2) go to a shop-ping mall or theater for relief, (3) if being out-of-doors is essential, find shady places to rest, (4) hold physical activities to a minimum during the earliest hours of the morning if necessary, (5) postpone usual outside chores, (6) wear loose, well-ventilated light-weight clothing, (7) drink large quantities of water, (8) refrain from excessive sun at the beach or pool, (9) shield the sun with a wide brim hat or umbrella, and (10) be alert to the "heat index."

The U.S. National Weather Service provides a heat index to indicate human discomfort associated with varying levels of relative humidity and temperature. Like the wind-chill index, it quantifies the humidity effect by factoring in the apparent temperature, solar radiation, wind speed, and barometric pressure on human comfort—a kin of "humidex."

As relative humidity increases, so does human discomfort. For ex-ample, at an air temperature of 90 degrees F (32 degrees C) and 50 per-cent relative humidity, the air feels as if it were 96 degrees F (36 degrees C). When the air is 95 degrees F and the relative humidity is 60 percent, the air feels like 114 degrees F. The reason is that as the air increasingly moistens and gets closer to saturation the resistance to moisture loss from the human body to the air also increases. People are more comfortable in cold air when humidity is higher, so humidifiers are often used in-doors in winter. Some people have come to rely on a humidistat in their home. It is an automatic device for controlling the extent the relative humidity is modified in a room.

MOST AND LEAST HUMID PLACES

In the United States the six cities having the lowest average annual relative humidity for July mornings (all below 30 percent) are: Reno, Sacramento, Boise, Phoenix, Salt Lake City, and Albuquerque. Cities re-cording the highest annual relative humidity on the average for July mornings (all above 89 percent) are: Mobile, Nashville, Jackson, New Orleans, Houston, and Charleston. The driest cities are all in the west with desertlike climates. The high relative humidity cities are located in the subtropical south.

The cities in the tropics record the highest average percentages of morning hour relative humidity. They are all well into the 90 percent range. Representative examples are: Kisangani, Zaire; Accra, Ghana; Abidjan, Ivory Coast; Kabale, Uganda; Belem, Brazil; and Kuala Lumpur, Malaysia. The lowest relative humidity percentages, worldwide, are all in desertlike climates. The are, by example: Aswan, Egypt; Khartoum, Sudan; Eilat, Israel; Islamabad, Pakistan; Jidda, Saudi Arabia; and Alice Springs, Australia. Their relative humidities are mostly in the 50 percent range on average.

People who travel should take into consideration the relative humidity at their point of destination. This will help them determine what kind of clothes to pack, or even if they want to go there at all. To illustrate two places with extremes in relative humidity—Aswan, Egypt, records a range of 15 to 31 percent relative humidity for afternoon averages throughout the year; Sena Madureira, Brazil, deep in the Amazon rainforest, has a super saturated relative humidity of 98 or 99 percent for every evening of every month.

THE LANGUAGES OF HUMIDITY

People over the years have created rather familiar analogies and expressions to describe the state of relative humidity when they report on the weather. Such terms as "stifling," "a wringer," "steambath," "sauna," "tropics," and "sponge" underscore high relative humidities. Low humidities with very high temperatures are frequently compared to such usage as a "furnace," "oven," "blistering," "scorcher," "Mojave Desert," "dry as toast," or "inferno."

HUMIDITY

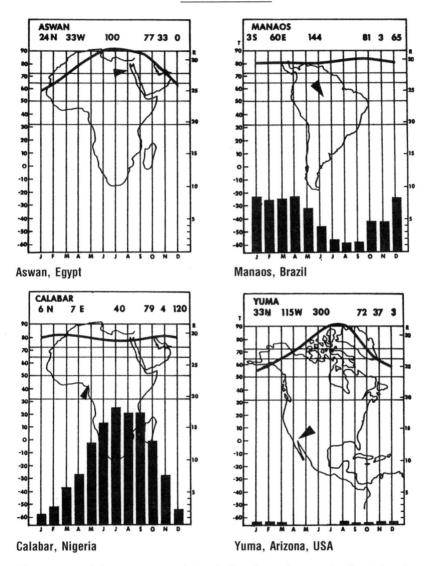

Aswan, Egypt

Manaos, Brazil

Calabar, Nigeria

Yuma, Arizona, USA

The amount of dampness or moisture in the air can have a significant bearing on the human body. It can affect energy levels, appearance, perspiration, breathing, and a person's general health, including their aging process. The accompanying climographs show wide disparities in rainfall amounts (see key in the Introduction), which lead to equally broad contrasts in wet and dry air contents at various locations.

SOURCES

Fisher, Robert Moore. *How to Know and Predict the Weather*. New York: New American Library, 1953.

National Disasters of North America (supplement). *National Geographic*. July 1998.

Nese, Jon M., and J. Grenci. *A World of Weather. Fundamentals of Meteorology*. Dubuque, Iowa: Kendall-Hunt Publishing, 1996.

Oliver, John E., and Rhodes W. Fairbridge, eds. *The Encyclopedia of Climatology*. New York: Van Nostrand Rheinhold, 1987.

Pearce, E. A., and C. G. Smith. *The World Weather Guide*. London: Hutchinson, 1984.

Petterssen, Suerre. *Introduction to Meteorology*. New York: McGraw-Hill, 1941.

Schneider, Stephen H. *Encyclopedia of Climate and Weather*. New York: Oxford University Press, 1996.

U.S. Department of Commerce. *Heat Wave: A Major Summer Killer*. Washington, D.C.: NOAA.

Walker, Sally M. *Water Up, Water Down: The Hydrologic Cycle*. Minneapolis: Carolrhoda Books, 1992.

White, C. Langdon, George T. Renner, and Henry J. Warman. *Geography: Factors and Concepts*. New York: Appleton-Century-Crofts, 1968.

Williams, Jack. *The Weather Book: An Easy-to-Understand Guide to the USA's Weather*. New York: USA Today and Vintage Books, 1992.

Yellin, Emily. "Heat-Related Deaths Rise in Several Cities." *New York Times*. July 28, 1999.

9

HURRICANES/TYPHOONS

The name hurricane is derived from the word *hurricane* of the Arawak speaking Indians of the West Indies. Basically hurricanes are inward spiraling storms with intense wind velocities (75 miles per hour or more), severe thunderstorms, and torrential rainfalls.

A misnomer is the act of applying a wrong name to some person or thing. Hurricanes and typhoons are violent tropical storms that are often given names that can sound casual or harmless. Such names as Andrew, Camille, Ginger, Flora, Bob, Gloria, Julio, or Hugo have been used to identify hurricanes in the past; however, their designations belie their behavior and the death and destruction they can bring. The annual selection of names is agreed upon in advance by the World Meteorological Organization of the United Nations. The nomenclature is multicultural. The Atlantic storms' list is mainly American and French. Central Pacific cyclones have Hawaiian names, and Hispanic names dominate storms of the eastern Pacific.

These powerful storms generally have a number of common characteristics. They form in late summer and early fall, have intertropical origins, and tend to start around the western margins of the ocean basins of the Northern Hemisphere. Usually there is an eerie calm in the air before a hurricane strikes. The hurricane itself is a huge round cone of counterclockwise churning air within a very low pressure system. The center, the eye of the storm, is calm. A hurricane can travel over the surface up to 100 miles per hour and extend to 300 miles wide. It can reach heights far above sea level.

Hurricane routeways include the West Indies, the Gulf Coast of Mexico and the United States, and the south Atlantic Coast of the United States. Hurricanes have, however, hit a number of states farther north, into Long Island and New England. In specific years the frequency has ranged irregularly from two to twenty-one hurricanes in the Atlantic with about an average of seven annually in the eastern Pacific. The hur-

Hurricanes are classified on a five-point scale based on the intensity of their wind speed, damage produced, and storm surge. Wind velocities range between 75 to 155 miles per hour. Gusts may reach 200 miles per hour. A typical high-category hurricane is deemed to be more powerful than a nuclear bomb. This scene shows the effect of the wind force of a lower-rated Florida hurricane on palm trees and fencing. (Courtesy of National Oceanic and Atmospheric Administration.)

ricane's "cousins," occurring in the northern Pacific Ocean, are called typhoons; those originating in the Indian Ocean are deemed cyclones. Fifteen percent of the world's population are at risk for experiencing a hurricane.

Hurricanes are now classified on a five-point Saffir-Simpson Scale based on the intensity of wind speed, damage produced, and storm surge. Thus category 1, winds between 74–95 miles an hour, causes minimal damage; a category 2 storm will destroy unsecured mobile homes and damage roofs, windows, and doors. Major hurricanes have winds over 110 miles an hour, and are classified as category 3, 4, or 5. Category 3 and 4 storms will cause moderate to significant structural damage. A category 5 hurricane, with winds over 155 miles an hour, will produce catastrophic damage.

As insurance claims have increased manyfold after each successive hurricane, some insurance companies have cut back sharply on sales in dwellings located on vulnerable coastal areas to those living along the eastern seaboard and Gulf of Mexico, from Maine to Texas. For home-owners living near hurricane-prone sites, insurance providers, in a move to check storm losses, have raised policy premiums considerably. Because of this, some people, unable to afford high-risk premiums for their coverage, remain uninsured and can lose everything should their home suffer severe damages during a hurricane.

AWARENESS, PREDICTION, AND PRECAUTIONARY MEASURES

An array of new high-tech procedures are being put in place to prepare people for the onslaught of potentially horrific hurricanes. Meteorologists at Hurricane Warning Centers are armed with Doppler weather radars and more tracking technology; and sophisticated computers and software, unknown a decade ago, help forecast hurricanes, their intensity, and paths. Specific hurricane alerts are broadcast with recommendations for evacuations, road condition reports, and what preparations households need to follow. Normal television and radio programs yield to emergency announcements. Telephone numbers are posted, offering all kinds of assistance. High-altitude jets loaded with new forecasting devices can now fly above, around, and into the eye of a hurricane for the purpose of assessing the strength of each storm and for collecting weather data. Each nine-hour hurricane flight, with a crew of up to a dozen, costs $22,500. Besides the reconnaissance aircraft geostationary satellites orbiting the earth at an altitude of 22,000 miles above the Equator provide imagery both day and night that is able to estimate the location, size, and intensity of a storm and its surrounding environment.

When a hurricane begins to threaten an area, police, firefighters, Red Cross workers, and sometimes State and National Guard units are placed on alert. People are encouraged to shutter or tape windows in advance in order to secure them and prevent breakage. They are advised to stock up on nonperishable food and drinking water, and to have a manual can opener, battery-operated radios, and spare batteries and flashlights. They are told to have rain gear on hand and make sure their medicines are in good supply. Disposable plates, glasses, and utensils, matches, sterno cans for cooking, candles or a lantern, and, if necessary, baby supplies, have to be ready for daily use. They need to bring lawn furniture and other loose, light objects inside the house. The family car should be ready with a full tank of gas, and of course, if the family has a pet, the animal's needs should be considered as well. It is wise to learn the location of official shelters in advance. And most importantly people in jeopardy of a hurricane must refrain from driving, unless ordered to do so, walking in flooded areas, or touching fallen high tension lines.

If one decides to remain at home and "tough it out" during a hurricane, it is a good idea to take some necessary precautions. Small appliances should be unplugged. It is also wise to fill the bath tub or a large container with water for emergency sanitary purposes should they be needed. The propane tanks need to be turned off. It is very advisable to stay away from windows, stay in an interior first-floor room such as a closet or bathroom, and to try to lie on the floor under a table.

After the storm passes listen to the local radio or television station to learn about safe road conditions, be alert to downed power lines, check all gas, water, and electrical lines for damage, and make sure the tap water is safe and not contaminated. It may have to be boiled to remove impurities. Stay clear of water accumulations, especially when using any kind of electrical tool or other outdoor utensil.

HISTORIC HURRICANES

A typical high category hurricane is many times more powerful than a nuclear bomb. Violent hurricanes in the United States have claimed many thousands of lives. Prior to the introduction of America's effective hurricane warning system, the number of deaths from these unmerciful storms was staggering: 6,000 were killed in Galveston, Texas (1900), Florida lost 1,836 (1928), and 600 were lost in a 1938 hurricane in the northeastern United States. From 1900 to 1994 it is estimated that there were 12,352 fatalities from hurricanes in the United States. The Florida hurricane of 1928 also killed 2,000 people in the West Indies and 300 in Puerto Rico. During the Galveston hurricane a twenty-foot wave of water suddenly engulfed the city, drowning hundreds. Half the homes were destroyed; property damage was enormous.

The exact movement of a hurricane is really unpredictable, often striking locations where people surmised they would be out of the line of its path. Hurricane Ginger, like an elusive speedy football halfback, roamed the North Atlantic for twenty-seven days, from September 10 to October 5, 1971, causing great displacement of people and a large number of insurance claims. Hurricane Camille, a two-day hurricane started initially along the coasts of Mississippi and Louisiana on August 17, 1969, reported gusts of wind to 200 miles an hour. Tides peaked at twenty feet above normal, and some areas received six to fifty inches of rain. Bridges, trees, homes, and entire towns were washed away. An estimated 258 people were killed, thousands injured, with over $1 billion in damages.

Hurricanes most often cause widespread torrential rains and flooding. Tropical storm Claudette (1979) brought forty-five inches of rain to an area near Alvin, Texas. Hurricane Diane (1955) brought floods to the northeastern sector of the United States that contributed nearly 200 deaths and $4.2 billion in damage. Hurricane Agnes (1972) produced floods in the same general region causing 122 deaths and $6.4 billion in damage.

Hurricane Andrew struck southern Florida, August 24–26, 1992, resulting in billions of dollars in losses for homeowners and insurance companies. Over 160,000 became homeless primarily from the relentless fierce winds that demolished thousands of homes. Afterward, in Andrew's wake, building codes there and elsewhere were raised to much higher minimum standards of construction, particularly to withstand high wind forces, which had proven ruinous.

Two historic fierce category 4 hurricanes struck the east coast of the United States almost exactly ten years apart during mid-September of 1989 and 1999. Hurricane Hugo, the first of the two, came ashore in downtown Charleston, South Carolina, killing thirteen people and costing that state more than $6 billion. Hurricane Floyd hit the border between the Carolinas and had far-reaching impacts. It was estimated that Floyd, which was 600 miles wide, with wind gusts up to 150 miles per hour, caused the evacuation of 2.5 million people from their homes. Tens of thousands of automobiles clogged the interstate highways. This despite the fact that multiple lanes were open only in an outflowing direction from the well-publicized, menacing approaching hurricane, causing a 100-mile backup of vehicles. Flooding occurred in most low-lying areas from Florida to New England amid torrential downpours of rainfall.

In Virginia, more than eighty navy ships with an estimated 30,000 sailors took to sea to escape the storm. Military planes had to fly to safe airfields. Commercial airlines throughout the country had to cancel hundreds of flights. Floyd was deemed to be one of the most disruptive storms of the century. In North Carolina, where the flooding lasted a week after Floyd's departure, major roads were blocked and cities were isolated.

Threats of starvation and disease hung over victims of Hurricane Pauline, southeast of Acapulco, Mexico, on October 8–9, 1997. Rescue workers were unable to reach 20,000 people due to washed out roads. Communities in the area reported going up to five days without food because helicopters were unable to bring relief supplies due to fog and rain. At least 240 people died as a result of the hurricane, and 300,000 were left homeless. Broken water systems and flooded sewer drains raised the specter of epidemics. Hundreds were quarantined; and emergency steps were taken to curtail the spread of cholera and prevent a crisis of people getting dengue, a painful mosquito-born fever and rash.

COPING AFTER THE HURRICANE

Coping with routine functions in the aftermath of a hurricane can be quite trying. Once the hurricane recedes, damages to the interior and the exterior of structures may require evaluation and cost assessments for repairs and replacements. An insurance adjuster must be called to evaluate losses and make repair estimates for policy coverage.

It is important to have all broken glass removed in and around the premises. All necessary household and pharmaceutical supplies must be replenished. Food items need to be checked for usability. In cases where the hurricane's impact has been especially traumatic, counseling of individuals may be recommended.

In the United States the government's Federal Emergency Management Agency (FEMA) constantly stands ready to come to the aid and relief of people victimized by a disaster, such as a hurricane. Teams of FEMA specialists are promptly dispatched to the distressed area with accompanying food, water, financial aid, rebuilding supplies, medical items, equipment for the restoration of the infrastructure, and health maintenance facilities. They are almost always assisted by the Red Cross, churches, social agencies, local authorities, and police and military personnel. FEMA has become one of the nation's most efficient and respected organizations.

HURRICANES/TYPHOONS

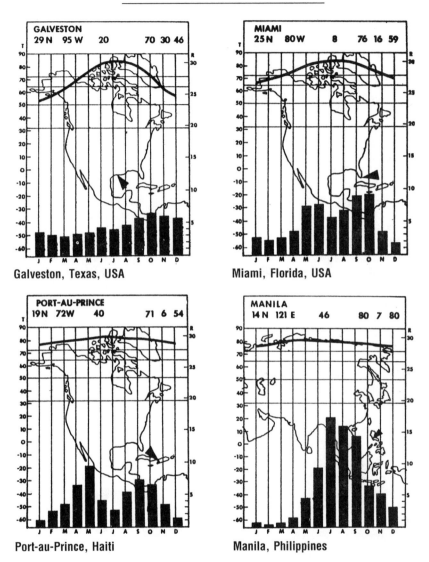

Galveston, Texas, USA

Miami, Florida, USA

Port-au-Prince, Haiti

Manila, Philippines

The residents of these cities have felt the wrath of numerous hurricanes in the past. They are situated along the perilous paths of periodic cyclonic storms having wind velocities in excess of 73 miles per hour, most often augmented by heavy rainfalls (see key in the Introduction). Hurricanes, wherever they may strike, have the capacity to bring death and havoc in their wakes. Even when people are warned in advance of an impending hurricane they are most difficult to avoid.

SOURCES USED

Battan, Louis J. *Fundamentals of Meteorology.* Englewood Cliffs, N.J.: Prentice-Hall, 1984.

Brindze, Ruth. *Hurricanes: Monster Storms from the Sea.* New York: Athenaeum, 1973.

Firestone, David. "Hurricane Aims at Coast of Carolinas." *New York Times.* September 16, 1999.

Fisher, Robert Moore. *How to Know and Predict the Weather.* New York: New American Library, 1953.

Halford, Ingrid. *Weather Facts and Feats.* (2nd ed.) Middlesex, Eng.: Guinness Superlatives Limited, 1982.

Henry, James A., Kenneth M. Portier, and Jan Coyne. *The Climate and Weather of Florida.* Sarasota, Fla.: Pineapple Press, 1994.

"Hundreds Dead or Missing as Hurricane Pauline Lashes Acapulco." *Home News and Tribune* (East Brunswick, N.J.). October 10, 1997.

"Hurricanes—Unleashing Nature's Fury." A Preparedness Guide. National Oceanic and Atmospheric Administration, National Weather Service, March 1996.

Mayfield, Max, and Miles Lawrence. "Atlantic Hurricanes." *Weatherwise.* February/March 1996.

"Mexican Hurricane Victims May Face Worse Travails." *Home News and Tribune* (East Brunswick, N.J.). October 19, 1997.

Moran, Joseph M., and Lewis W. Morgan. *Essentials of Weather.* Englewood Cliffs, N.J.: Prentice-Hall, 1995.

Natural Disasters of North America (supplement). *National Geographic.* July 1998.

Rosenfeld, Jeff. "The Forgotten Hurricane." *Weatherwise.* August/September, 1993.

Schneider, Stephen H., ed. *Encyclopedia of Climate and Weather.* New York: Oxford University Press, 1996.

Simpson, Robert H., and Herbert Riehl. *The Hurricane and Its Impact.* Baton Rogue, La.: Louisiana State University Press, 1981.

Smothers, Ronald. "Hurricane Slams Into Florida With Winds of 145 M.P.H." *New York Times.* October 5, 1995.

Statistical Abstract of the United States 1998. (118th ed.) Washington, D.C.: U.S. Bureau of the Census, 1998.

Tannehill, I. R. *Hurricanes, Their Nature and History.* Princeton, N.J.: Princeton University Press, 1938.

U.S. Department of Commerce. *Hurricanes, The Greatest Storms on Earth.* Washington, D.C.: NOAA, 1994.

10

ICE

Ice is nice—except when it is brought on by a winter storm. An ice storm has the potential to leave homes without power, heat, running water, and even cause structural damage to the building. It can make roads slick and treacherous and generally wreak havoc on the natural environment. Particularly perilous is the toll an ice storm might take in causing deaths and injuries, especially those due to falls.

Ice is the solid form of water. It can form in snow, on water bodies, and as a key composition in glaciers. Water vapor freezing on trees, vines, wires, and all kinds of objects may resemble a glass coating called *glaze ice*. Hanging icicles can look like stalactites that might be fixed to the roof of a cave. Ice is formed when the water moisture in the air is pure, and the temperature is 32 degrees F (0 degrees C). Impure water, such as seawater, which has salt content, freezes at 27 degrees F. When water freezes, it increases in volume by one-eleventh. That is why water pipes can split open as temperatures rise above freezing unless insulated or warmed. And for the same reason, anti-freeze or alcohol needs to be added to automobile radiators to prevent serious damage to the cooling system of the engine.

SLEET AND FREEZING RAIN

Sleet is generally considered to be frozen rain or rain that freezes into ice pellets during its fall. It usually bounces when hitting a surface and does not stick to objects. It could also be a mixture of rain with hail or snow. Both can place an icy coating on many kinds of manmade or natural surface features and cause a hazard to motorists.

HISTORIC ICE STORMS

Man can produce ice and prevent ice from forming on a very limited scale. However, Mother Nature's powerful ice-making machinery is

Guests use glasses made of ice at the bar, which is made of ice, in the Jukkasjaervi Ice Hotel in northern Sweden in this January 1997 photo. The whole hotel is made of snow and ice. Over shots of vodka—there's no beer because its low alcohol level means it would freeze in the 23-degree room—they talk bravely of the night to come. (*Credit:* AP/Wide World Photos. Reprinted with permission.)

much too awesome to stop. She has left her icy calling card with the people of vulnerable sections of Europe and North America in the mode of ice and weighty ice storms for each and every ice-forming season, which could last for many fall and winter months at higher latitudes. In early February 1951, one of the greatest ice storms in U.S. history, in terms of its expanse, produced an extensive coating of ice up to four inches thick from Texas to Pennsylvania. As is common in such storms, icy roads were closed; absenteeism from work and school was way above the norm, powerlines were downed causing outages, and damage was in the millions of dollars. The total number of deaths reached twenty-five during the weeklong storm. The state of Tennessee was the hardest hit. A similar ice storm on November 24, 1997, spread freezing rain and ice over the southern Great Plains from Texas to Missouri. The weather was blamed for seventeen deaths.

Early winter ice storms can arrive unpredictably on cattle ranches before precautions can be taken. A storm like this once swept through Dexter, New Mexico, killing 3,000 calves on one ranch, causing $500,000 worth of losses to its owner.

Ice storms know no boundaries. Over a five-day period, January 5–10, 1998, large areas of northeastern United States and Quebec, Canada, were paralyzed by a devastating ice storm. It was the biggest disaster ever to hit the state of Maine. Millions of people were affected, including three million Canadians. In upper New York, 2,288 National Guard soldiers were dispatched to aid the victims of the storm; and in eastern Ontario and Quebec, 15,000 Canadian troops were deployed to offer relief and restoration of normal services. In some locations it took two weeks to resume usual operations.

Thousands throughout the area were taken to shelters where cots, blankets, meals, and water were available. At least nineteen deaths were blamed on the storm. Losses totaled $1.5 billion. Maple syrup farmers lost prize groves. New York dairy farmers were forced to dump 366,000 gallons a day of fresh milk, unable to run milking machines or get their products to processing plants. Power outages caused considerable spoilage to cheese plants and fruit storage facilities. Household food supplies were frequently kept cold by cutting large blocks of ice and inserting them into powerless refrigerators. People took safety risks by using outside propane gas grills in their homes to cook and keep warm. A number of fires resulted. Thousands of homes were flooded when frozen pipes burst.

ICEBERGS

Icebergs, once the dread of sailors, are broken-off ends of glaciers that slide into the sea. Those that float in the north Atlantic Ocean come from the island off Greenland. They may vary in size from small to mountains of ice a mile across and more than 200 feet above the water. The part below the water is about seven times as large as the portion visible above the water's surface. This more than likely led to the origin of the expression "tip of the iceberg," implying only a portion of an object being visible.

An iceberg the size of Rhode Island that sheared off the coast of Antarctica in May 1996 covered 1,400 square miles when it split off. The scientists that spotted it from their research ship originally estimated it to measure about 54 by 27 miles with 160-foot walls. They predicted it could drift for ten years before melting. A few hydrographers, scientists that study water bodies, maintain that a gigantic iceberg can now be profitably towed by powerful tugboats from cold source regions and deposited along the coasts of deserts in order to serve as a reservoir or water supply lake.

DANGERS OF ICEBERGS

Icebergs can be extremely dangerous to commercial and sport fishing craft. Many boating casualties have left the seamen of the Grand Banks south of Newfoundland, Nova Scotia, New Brunswick, and Prince Edward Island waters with a watchful eye looking for drift ice and icebergs. They are well aware of how an unseen iceberg, if struck by a boat, can tear open its hull. One of the greatest sea disasters in the world's history was the sinking of the *Titanic* on the night of April 15, 1912. The 46,000-ton luxury liner, the largest ship afloat at that time, struck an iceberg off Newfoundland causing 1,517 persons to lose their lives as the ship went down. Now radar equipment is used by vessels to warn of iceberg positions ahead.

USES OF ICE

Before ice could be machine-made, ice had an important use in people's lives. Wherever and whenever possible, people harvested ice from nearby frozen lakes and rivers. They usually stored it underground in ice cellars. Some ice was kept in wooden icehouses. Blocks cut in twenty-one-inch squares were kept cold by blanketing them with layers of straw until they were needed during the warmer months.

Despite the fact that ice conditions can be fraught with potentially dangerous situations, it can also provide many opportunities for wholesome fun-filled activities. Ice festivals, family skating parties, sport competitions, and figure skating have grown in popularity. There is even an interest-arousing rise in creative ice sculpturing on some college campuses.

A world-famous snow festival takes place each February that attracts a quarter of a million people to Sapparo, a city located on Japan's northern island of Hokkaido. During the festival the city is virtually transformed into outdoor museums of sculptured snow and ice.

Visitors to Sapparo may marvel at the displays of 65 foot ice sculptures or gigantic snow castles made from tons of snow trucked in from the nearby mountains. Snow seems to abound for many people experiencing their very first playful encounter with it. There are igloos to explore, ice and ski slopes to slide down and even an ice stage for musical performers and other kinds of entertainers. The Sapparo festival has been lauded for the past fifty years for its wholesome family-fun appeal.

The most unique example of ice usage can be found in Jukkasjaervi, Sweden, a hamlet located 100 miles north of the Arctic circle where a one-story twenty-room hotel is built out of ice every December and lasts until sometime in May. The Inshotellet, as it is known, played host to about 4,000 people at a room rate of $75 per night in 1997. Guests wear snow suits, sit on ice blocks covered with reindeer skins, employ candles for heat and light, and rely on sleeping bags for a cozy quiet night of sleep, even when the temperature outside is −8 degrees F. The igloo-style "Holiday Inn" is constructed of five-foot-thick packed snow and ice blocks sawed from a nearby river. No beer is served in the hotel because it would freeze in the 23-degree room; however, a smiling maid brings hot lemonade to each guest's room every morning and reminds them that the nearby sauna and bathrooms are open—outside.

FROZEN BODIES OF WATER

Ice formed over water bodies can be extremely dangerous for people, especially for unsuspecting children unaware of the hazards associated with seemingly frozen surfaces. Each winter hundreds of people become trapped while fishing on ice floes and need to be rescued by police boats and military helicopters. Many ice fishing enthusiasts merely fall through the ice and, unable to get help, drown; or if pulled out suffer from hypothermia, a subnormal body temperature condition that can be fatal. Snowmobilers in increasing numbers are dying or injuring themselves by driving across dangerously thin ice. Iceskaters who fail to heed warnings that prohibit venturing onto unsafe ice ponds place themselves at a high risk of drowning. Young children playing unsupervised ice hockey on what they may deem to be thick ice could, in reality, be skating over thin ice with deep water below. Iceboats skimming over frozen lakes, thought to be solid can easily result in a tragedy for those aboard if the ice fails to hold the weight of the boat and its sailors.

ICE

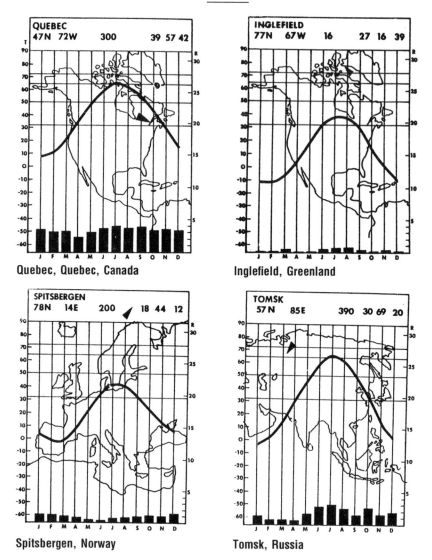

QUEBEC
47N 72W 300 39 57 42

Quebec, Quebec, Canada

INGLEFIELD
77N 67W 16 27 16 39

Inglefield, Greenland

SPITSBERGEN
78N 14E 200 18 44 12

Spitsbergen, Norway

TOMSK
57N 85E 390 30 69 20

Tomsk, Russia

Ice conditions can be fraught with potentially dangerous outcomes. Yet, the formation of ice can also provide positive benefits and opportunities for fun-filled wholesome festivities and activities. People living in ice-prone regions tend to know how to adapt to its hazards, yet many welcome its arrival as a natural friend rather than an unwanted foe. As depicted by the temperature curve (see key in the Introduction), the residents of these climograph places experience icy conditions for months on end.

SOURCES

Cole, Franklyn. *Introduction to Meteorology*. New York: Wiley, 1980.

"Gigantic Iceberg May Drift for Decade." *Sun-Sentinel* (Ft. Lauderdale, Fla.). November 16, 1996.

Goldberg, Carey. "In Maine, Ice Shatters Trees and Crust." *New York Times*. January 13, 1998.

Hare, F. K., and M. K. Thomas. *Climate Canada*. New York: Wiley, 1974.

Heintz, Jim. "Cold Night's Sleep at Ice Hotel." *Sun-Sentinel* (Ft. Lauderdale, Fla.). February 12, 1998.

"Hundreds of Canadians Are Trapped on Ice Floes." *New York Times*. January 27, 1997.

"Ice Storm Cost Military an Extra $41 Million." *Sun-Sentinel* (Ft. Lauderdale, Fla.). January 13, 1998.

"Ice Storms Blamed for 17 Deaths as Roads in Plains Become Slick." *New York Times*. November 26, 1996.

"Icy Storm Kills 3,000 Calves on a Ranch in New Mexico." *New York Times*. February 26, 1997.

Kristof, Nicholas. "Ice, Ice, Baby." *Sun-Sentinel* (Ft. Lauderdale, Fla.). January 9, 2000.

McDonald, Kim A. "Preserving a Priceless Library of Ice." *Chronicle of Higher Education*. August 1996.

"Mammoth Rigs on the Ice, Holding Their Breath." *New York Times*. March 22, 1999.

Marshall, Roxy. "Ice-Storm Knocks Out Power." *Sun-Sentinel* (Ft. Lauderdale, Fla.). December 28, 1996.

National Disasters of North America (supplement). *National Geographic*. July 1998.

"Northeast and Canada Reeling from Storms." *Sun-Sentinel* (Ft. Lauderdale, Fla.). January 10, 1998.

Rimer, Sara. "Crackle of Icy Lakes Means It's Spring." *New York Times*. May 8, 1997.

"*Titanic* Sinks Four Hours After Hitting Iceberg, 866 Rescued, Probably 1250 Perish." *New York Times*. April 16, 1912.

Yardley, Jim. "Northern New York Battles Ice Damage." *New York Times*. January 12, 1998.

A dry field of brush or forest can readily become a tinderbox for lightning strikes. Here firefighters take a well-deserved rest and refreshment break from fighting a very troublesome forest fire in Shasta County, California, August 1992. (Courtesy of American Red Cross. Photo by Maggie Hallahan.)

11

LIGHTNING/THUNDER

The sights and sounds of lightning and thunder are somewhat like a Fourth of July fireworks display. Like an elaborate theatrical production, they can be spectacular to watch and explosive to listen to. Since lightening could be life threatening, it should be viewed with caution, whenever possible under cover of shelter, for the sake of safety.

Lightning is in simple terms a flash of light in the sky caused by the discharge of atmospheric electricity from one cloud to another or between a cloud and earth. Lightning can strike the earth in the form of a single stroke, a bolt accompanied by the sound of thunder, with forks or branches from the main channel, or in the form of sheet lightning, a more common blanketlike illumination that tends to diffuse over the entire sky. Warm summer nights favor the occurrence of heat lightning and thunder, the sound that follows a flash of lightning caused by the sudden heating and expansion of air by electrical discharge.

Storm watchers can estimate the nearness of a storm by counting the seconds between the time the flash of lightning is sighted and the ear-shattering sound of a thunder roll. Every five-second difference indicates a distance of one mile. In as much as storms usually move at a speed 25 to 35 miles per hour, that is a time-needed rule-of-thumb to guide those seeking shelter.

LIGHTNING/THUNDER IN MYTHS AND FOLKLORE

Long before there was an accepted explanation for the causes of lightning and thunder, the folklore of every culture held a wide array of beliefs and reasons for this natural phenomenon. Early peoples concluded that it was an act of warning against evil by the deities. In the Orient it was the work of the dragons. In Old Testament times lightning struck fear in the hearts of the people, who justified it as a divine will. Christians thought it to be a sign of probable punishment for sin. Thunder was considered to be Thor throwing his large hammer at giants, God chasing the devil, or American Indian's notion that the thunderous noise from above was the flapping of the wings of huge "thunder-birds." Later superstitious practices included the idea that fires started by lightning could only be put out by milk; trees felled by lightning should never be used as fuel, but were permissible for use as a magical medicine to cure ills.

Certain animals such as cats, dogs, mules, and horses were said to attract lightning. Holding specific objects like the Bible, candles, bells, and salt would either prevent a lightning strike or make it subside. In the house, featherbeds were believed to offer a sure refuge, while occupants were warned to stay away from windows, mirrors, and chimneys. People were expected to remain absolutely quiet during thunder because it was thought that talking could be interpreted as ridiculing the loud booming noises in the sky and provoke a punishing lightning bolt upon the violator.

Some weird legends handed down by early settlers in the Hudson Valley of New York attributed spoiled milk to the rumbling sounds of thunder. Baldness was also blamed on thunder. The ghosts of explorer Henry Hudson and his crew from his *Halfmoon* boat were thought to be playing ninepins in the Catskill Mountains where the rolling sounds seemed to originate.

Some lightning myths still persist. Among them is the belief that if it is not raining there is no danger from lightning. Another is the myth that rubber soles or shoes or rubber tires on a car will protect a person from being struck by lightning. Or that people struck by lightning carry an electrical charge and should not be touched. Also erroneous is the notion that "heat lightning" occurs after very hot summer days and poses no threat.

DANGERS OF LIGHTNING

For some people the experience of hearing the sounds of thunder can be frightening. Witnessing a lightning storm can cause fears that can last a lifetime. Some demonstrate their fears in unusual ways. They cover their eyes and ears, cry, pray out loud, scream, want to be held, or even hide in closets until the storm abates. Children are especially fearful of these kinds of electrical storms. Animals tend to behave in very odd and unpredictable ways during periods of lightning and thunder.

Lightning can be very dangerous. It occurs with all thunderstorms. The typical thunderstorm is fifteen miles in diameter and lasts an average of thirty minutes. Lightning strikes 400 people per year in the United States and on average takes ninety-three lives nationwide. It is reported to be the most severe killer from a weather cause. In a thirty-four-year period in the United States, lightning killed about 7,000 people, 55 percent more than deaths due to tornadoes over the same period. It has been estimated that lightning strikes the ground 100–125 times per minute on a global basis. The current of a lightning discharge averages around 30,000 amperes. One-tenth of an ampere can kill a human. The temperature inside a lightning channel reaches 50,000 to 60,000 degrees F, several times the heat of the sun's surface. Contrary to popular belief lightning can strike in the same place many times. Tall buildings, such as the Empire State Building in New York, are particularly vulnerable. There lightning has struck as many as twelve times in twenty minutes and as often as 500 times a year. That is why observation decks of skyscrapers are closed to visitors during lightning strike periods.

Lightning takes the path of the least resistance in its path to the ground, so standing under a tree while seeking shelter in an electrical storm is very unwise. Lightning can travel into the roots of a tree as well, even into tents where campers or soldiers falsely feel they are shielded from its wrath. Many, particularly golfers playing out in the open, run a high risk of being struck by a lightning bolt. In fact most lightning deaths and injuries occur when people are caught outdoors. That is why some golf courses in Florida and elsewhere now sound a loud fire siren mandating that all players leave the golf course immediately upon the threat of an imminent lightning strike.

It is not unusual for people on beaches, in swimming pools, on fishing piers, or in small boats, to be ordered to get out or away from the water. An open beach umbrella is hardly safe. Lightning often strikes a conductor of electricity, such as water or metal. Metal roof sheds invite tragedy for those that enter them for shelter unless they have a lightning rod placed high on the building. The rod, acting as a conductor, is grounded at the lower end and diverts lightning from the structure. People are advised to refrain from using the telephone during lightning because telephone lines can carry the electrical charge right to a person holding the receiver, resulting in death by lightning within a home. And most damaging of all are the fires that start in forests, brush, and buildings that are caused by lightning strikes when there are tinderbox drought conditions on land. Though quite rare, lightning can cause airplane crashes as it did years ago at Elkton, Maryland, killing eighty-two passengers and crew. Being inside an automobile, on the other hand, is very safe.

HIGH-RISK AREAS

Lightning is a real risk for people living in Florida, which leads the nation in deaths and injuries from lightning strikes. Lightning is so common there during the rainy summer months that everyone seems to be conditioned to periodic delays in the completion of sporting events, swim parties, picnics, and at outside construction sites. Lightning detection systems have been installed in many buildings to warn of conditions right for strikes. Still, in the state of Florida, at least eight people died in 1997 and more than fifty-five were reported injured by lightning, which amounted to about 10 percent of the nation's casualties that year. A study of Florida lightning deaths and injuries revealed that the likely targets were 25 percent in, near, or around water, 22 percent under or near trees, and 7 percent during leisure which included golf.

Except for the polar regions and infrequently in the higher latitudes close to the poles, thunder is heard around the world. The sound of thunder can be heard seven miles away. Bogor, Java, is known as the most thunderous place of all. It holds the record for 322 days of thunder within a year. There is little wonder about thunder in that location. Central Africa has 150 days of thunder; Panama, 135; southern Mexico, 142; and central Brazil with 106. Central Europe and Asia average less than twenty thunder days per year.

In the United States most thunder days occur in Florida, along the Gulf of Mexico, and the mountainous areas of New Mexico. Tampa, Florida, averages ninety-four thunderstorms per year, more than any other U.S. spot. Santa Fe, New Mexico, is second with seventy-three. The West Coast of the United States receives only one to four thunderstorms per year on average. Each moment there are 1,500 to 2,000 thunderstorms taking place at the same time. There are approximately sixteen million thunderstorms that take place around the world every year.

HISTORIC INCIDENCES OF LIGHTNING STRIKES

Some bizarre cases of nearly unbelievable lightning strike stories have actually taken place. On June 28, 1975, famous golf champion Lee Trevino and two other golfers were struck by lightning at the Western Open Golf Tournament in Oak Brook, Illinois. A golfer in Memphis, Tennessee, was hit by a lightning bolt. It passed through his neck, down his spine, exited his body through a pocket holding his keys, then struck a nearby tree. Amazingly, he survived. A gruesome result of lightning happened on May 16, 1984, to a farmer near Buckeye, Arizona. The farmer was killed when the charge struck at the top of his head, then went through his body to his left heel, causing his boot to explode. His watch and trouser zipper melted. The following year lightning struck trees about 150 yards from a home in Alabama and followed the driveway to the house. The charge entered the house, burned all the electrical outlets, ruined the appliances, and blasted a hole in the concrete floor of the basement.

LIGHTNING/THUNDER

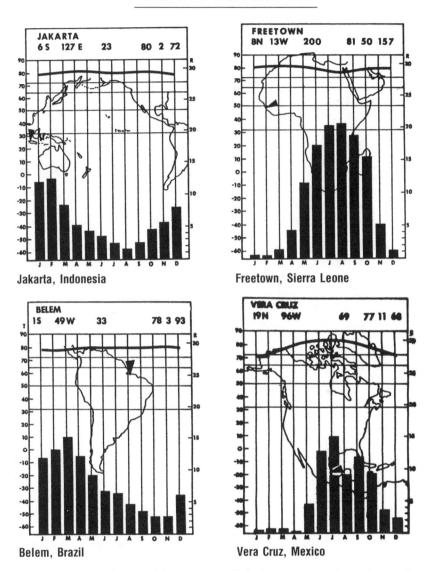

Jakarta, Indonesia

Freetown, Sierra Leone

Belem, Brazil

Vera Cruz, Mexico

The experience of a frightful thunder and lightning storm can be so traumatic that its impact may last for a lifetime. It is a well-known fact that lightning strikes can be exceedingly dangerous, yet foolishly some people take unwise risks by not applying safety precautions during an electrical storm. Lightning is reported to be the greatest cause of death from all weather-related disasters. People living in these climograph cities, whose rainfall curves depict heavy storms (see key in the Introduction), are confronted with frequent lightning and thunderstorms.

SOURCES

Battan, Louis J. *Weather*. Englewood Cliffs, N.J.: Prentice-Hall, 1985.

Browne, Malcolm W. "Are Lightning Balls Spheres of Plasma?" *New York Times*. September 10, 1996.

Fisher, Robert Moore. *How to Know and Predict the Weather*. New York: New American Library, 1953.

Kramer, Stephen. *Lightning*. Minneapolis: Carolrhoda Books, 1992.

Lutgens, Frederick, and Edward Tarbuck. *The Atmosphere: An Introduction to Meteorology*. Englewood Cliffs, N.J.: Prentice-Hall, 1989.

Mireya, Navarro. "When Lightning Strikes, Lives Are Changed." *New York Times*. September 1, 1998.

Natural Disasters of North America (supplement). *National Geographic*. July 1998.

Pearce, E. A., and Gordon Smith. *The Times Books World Weather Guide*. New York: Times Books, 1984.

"Relief, Hope from the Sky: Rain Helps Against Fires." *Sun-Sentinel* (Ft. Lauderdale, Fla.). December 6, 1997.

Salanane, L. E. *Lightning and Its Spectrum*. Tucson: University of Arizona Press, 1980.

Schneider, Stephen H., ed. *Encyclopedia of Climate and Weather*. New York: Oxford University Press, 1996.

"Thunderstorms and Lightning—The Underrated Killers." A Preparedness Guide. National Oceanic and Atmospheric Administration. National Weather Service. January 1994.

Uman, Martin A. *Lightning*. Mineola, N.Y.: Dover, 1969.

12

MONSOONS

Of all the different climates found in the world the monsoon climate has the most defined characteristics. The word monsoon derives from an Arabic word for "season." It is broadly applied to all wind systems that seasonally reverse their prevailing winds on a continent. The best examples of these opposite direction wind flow patterns are found in Southeast Asia, from the Arabian Sea to China, the Gulf Coast of the United States parts of Africa, and northern Australia. More than anywhere else the subcontinent of India is greatly impacted by the monsoon. With few exceptions the people of the monsoon regions have become conditioned to rhythmic rainy summers and dry winters. Though both kinds of weather are predicted, it is the arrival time, duration of stay, and the extent of the rainfall or drought that is uncertain. That is why the monsoon can be so whimsical, and that behavior can prove to be a matter of life or death for monsoon region inhabitants where famine is hardly unknown. More than half of the world's people live in monsoon Asia, many of them in extreme poverty.

The decisive difference between the two seasons is based on a fundamental meteorological fact. Land heats and cools more quickly than water. Cooler heavier air tends to displace the warmer lighter air and push it skyward. Therefore, the monsoon effect brings denser moisture-laden air from the sea in summer (wet monsoon) and conversely causes the dry air over the land to flow seaward during the winter months (dry monsoon). In a way the monsoon's disposition is like a person who displays periodic mood swings or a double personality. Almost 80 percent of India's rainfall comes during one season alone, the summer months from mid-June though September, and has a tremendous impact on the lives of the people, especially those trying to survive in rural villages. Where the brutal summer heat and drought parches the earth amid stifling dust storms, people are encouraged to wear protective face masks in order to filter out lung-clogging dust particles.

A boy carries drinking water through a flooded street in Calcutta, India in September 1999. Heavy monsoon rain lashed India's West Bengal state. Monsoons such as this one can cause house collapses, which can kill people. (*Credit:* AP/Wide World Photos. Reprinted with permission.)

In the cities, the seasonal climate records also tell a revealing story. When summer temperatures are at their highest levels, the amount of precipitation is staggering compared to the slightly cooler months of autumn and winter. From January to May only one inch of rain falls in total. From October to December the amount is a meager 3.1 inches. However, from June to September the rainfall measures 67.2 inches. In Dhubri, India, the winter and fall months bring only 13 percent of the city's 97 inches of rain; and in Darjeeling, in the northern mountains, 111 inches of the city's total rainfall of 120 inches arrives from May to October.

Half of India's arable land depends solely on monsoon rains and a single summer growing season. The rain is usually very heavy, sheetlike, with wave after wave that can readily turn the earth into a swamp or mud field. Wells fill up and lakes overflow. Flimsy mud huts disintegrate and rural roads become brownish colored mini-rivers. The thirsty brown rice fields are suddenly transformed into emerald green productive lands with the arrival of the monsoon. Across the plains of northern India and Bangladesh, entire villages are sometimes overflowed by monsoon rains causing all those residing there to flee to higher ground. In those areas where heavy rains ram against mountains, roads become blocked by muddy landslides. Often entire villages are wiped out with considerable loss of life.

EFFECTS OF THE MONSOON ON DAILY ACTIVITY

The monsoons determine many essential activities for the people of South Asia and much of the Orient. Despite some use of irrigation, the arrival of the monsoon triggers a burst in rapid farming operations. In the hot dry season things are apt to be rather dormant for the peasants. With the intense heat and very arid soil conditions during the months preceding the monsoon, little planting or farming tasks can be done. With the coming of rain, work accelerates because the growing season has to be compressed into a short period of time. There is little time for cooking or anything but work. All hands, including women and children, are out in the rice field, plowing, planting, and harvesting.

HISTORIC FLOODS CAUSED BY MONSOONS

For two months during the late summer of 1998 the longest lasting flood in the history of Bangladesh caused twenty million people of that country to evacuate their homes. Many died. It was estimated that 3.5 million Bangladeshis developed serious illnesses. The monsoonal effect has caused numerous memorable floods and tidal waves throughout much of South Asia during the past half century. In 1955 as many as 1,700 people perished in India and Pakistan, 10,000 died in Bangladesh during 1960, and an additional 780 people lost their lives in northeast India in 1968. Floods in the Himalayan region of India in 1970 cost 500 more lives. From June to September 1978 there were 1,200 victims of flooding brought on by the monsoon summer rains throughout north India. On August 11, 1979, an incredible total of 15,000 people living in and around Morvi, India, suffered violent deaths due to the extraordinary rains there. Again, in 1987 rampant floods visited upon Bangladesh caused 1,000 to die, and in 1988 a similar number of north region Indians suffered the same fate.

MONSOONS IN MYTH AND FOLKLORE

In most of rural Asia, monsoon signifies the renewal of life. Indian girls celebrate the return of the monsoon and the promise of prosperity by singing ancient melodies and reciting poetry that imply that winter sun brings misery and dark clouds symbolize rain and happiness. It is said that a beautiful woman has hair black as monsoon clouds and eyes that flash like lightning, a positive sign of a welcomed thunderstorm. In Nepal, women gather in the rivers for a ceremonial bath rejoicing in the arrival of life-granting rains and to honor the Hindu god who makes sure that all streams manage to cascade down the mountains onto the plains below. Almost everywhere, when the rains come, people run into the street, their arms spread pointing to the sky in a gesture of gratitude.

MAJOR PROBLEMS IN THE CITIES

The powerful monsoon rains hammer the cities of Asia as well. The lack of storm sewers promote flooding of the worst kind. In some Indian cities located along river banks, floods mixed with raw sewage, abandoned vehicles, and pull carts, rotten grain and floating carcasses of animals turn city streets into nightmarish disaster areas, unfit for human traffic.

Across India low reserves in reservoirs, before the coming of the monsoon rains, means that electric power can be rationed or cut off for consumers. In the past, movie theaters were only allowed one showing a day, neon display lights were turned off, stores were compelled to close at sundown, and schools closed due to the lack of drinking water for consumption and sanitation. About half of the nation's electricity is generated by hydropower, and thus by the monsoons. In India, society can function with a limited electricity supply, but drinking water in short supply can threaten the lives of the less fortunate who are unable to obtain it. That is why they covet and search for water during long dry spells, which can lead to fighting and even looting, and why a water vendor, pulling a barrel of water on a wooden cart through a city slum district, can easily sell water for only one rupee. But, once the monsoon visits the city, his sales "dry up" almost at once.

MONSOONS

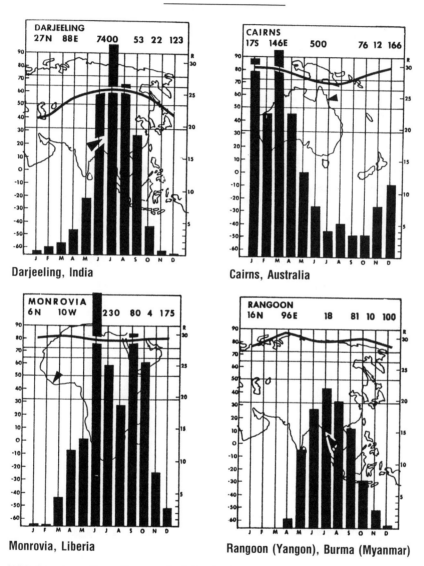

Darjeeling, India

Cairns, Australia

Monrovia, Liberia

Rangoon (Yangon), Burma (Myanmar)

With few exceptions the people living in the monsoon regions, as indicated by these representative climographs, where the rainfall curves at some points are off the charts (see key in the Introduction), have become highly conditioned to rhythmic rainy summers and dry winters. The very nature of their lives seems to be guided, if not controlled, by this decisive "weather calendar." More than half of the world's population lives in monsoon Asia, many of them in extreme poverty.

SOURCES

Blair, Thomas A. *Climatology; General and Regional*. New York: Prentice-Hall, 1942.

Burns, John F. "Toll of Pilgrims in Kashmir Storm Put at 160." *New York Times*. August 26, 1996.

Dugger, Celia W. "Monsoon Hangs On, Swamping Bangladesh." *New York Times*. September 7, 1998.

Farah, Mounir A., et al. *Global Insights*. New York: Glencoe, Macmillan/McGraw-Hill, 1994.

"Killer Storm Headed to Bangladesh." *New York Times*. June 6, 1996.

Oliver, John E., and Rhodes W. Fairbridge, eds. *The Encyclopedia of Climatology*. New York: Van Nostrand Rheinhold, 1987.

Pearce, E. A., and C. G. Smith. *The World Weather Guide*. London: Hutchinson, 1984.

Schneider, Stephen H., ed. *Encyclopedia of Climate and Weather*. New York: Oxford University Press, 1996.

The World Almanac and Book of Facts. Mahwah, N.J.: Primedia Reference, 1999.

13

MUDSLIDES

Mud—the wet, soft, sticky earth almost always connotes something negative. It can slow or halt vehicles, soil objects, cloud waters, contribute to the outcome of military battles, or most notably, alter the quality performances of a sporting event. It can also injure and kill people, damage property, close roads, and bring untold sudden misery to communities the world over. The forces of weather, such as torrential rains, flash flooding, melting snow and ice, aided by hilly or mountainous terrain, can combine to cause vicious mudslides of disastrous proportions.

HISTORIC MUDSLIDES

At Choco, a remote town 400 miles southeast of Lima, Peru, a landslide caused by heavy rains obliterated everything in sight, killing dozens of people. About sixty people from the town were out of danger when the avalanche hit because they were attending Mass in a church on a nearby hill.

Hillside towns in Bolivia, Colombia, and Brazil have also experienced mudslides in which death counts have caused great grief and anguish. In addition, many thousands of poor farmers became homeless. In 1992 more than 200 miners were killed when part of a mountainside slid over a mine sixty miles north of LaPaz. Six years later a gold mine in the same general area was buried by tons of muck, mud, and rocks causing the deaths of 140 miners and seriously injuring many more as they slept in their huts.

On May 16, 1998, towns at the base of Mt. Sarno east of Naples, Italy, inundated by rivers of mud and torrential rains, took the lives of 400 residents. More than 1,500 people lost their homes. Experts blamed the disaster on forest fires in the area that destroyed vegetation that would have kept the sodden earth from slipping. Also at fault was the construction of illegally built homes that did not meet required building codes.

The devastating effects of mudslides as shown in this picture taken along the Pacific Coast Highway in California in 1998. This home and at least one other at the top of the slide were scheduled for demolition at the time. (*Credit:* AP/Wide World Photos. Reprinted with permission.)

In March 1998 rows of oceanfront houses gouged out of the cliffs along the Pacific Coast dangled on the edge of crumbling foundations of rock and sand. For some home owners stepping just a few feet out the back door would mean a seventy-foot plunge into a steep abyss below.

In much of Central America during the first week of November 1998 there was absolute despair across a 500-mile corridor of hillside cities and villages from Nicaragua and Honduras into southern Mexico. The gloom and hopelessness that permeated that region at that time and the weeks that followed were brought on by Hurricane Mitch, one of the worst hurricanes of the twentieth century. The devastating four-day storm triggered massive floods and huge mudslides that suddenly wiped out practically every dwelling in its unmerciful path, resulting in a million people being left homeless. The casualties were staggering. An estimated 11,000 died, and 6,500 were reported missing, along with billions of dollars in damage that would be impossible to recover for decades.

The "sea of mud," as it was described by survivors, buried so many people in Nicaragua that parts of that unfortunate nation were morbidly referred to as a "national cemetery." The president of Nicaragua likened the disaster to a "panorama of death, desolation and ruin throughout the entire country." Compounding the pain and suffering of the survivors was the destruction of 75 percent of the region's infrastructure. The ruin of roads, bridges, and communication systems prevented rescue attempts and supplies from reaching thousands of people cut off by the flooding and inundating blankets of mud. That very frustrating situation led to further deaths, infection, spreading disease, and even starvation, despite the valiant efforts of many worldwide relief agencies to aid the victims.

A torrent of mud and boulders loosened by rain rushed down a hillside and buried a hamlet in the Amazon jungle during January 1999. Fifteen people were killed and four were reported missing at Jazan, 4,600 miles north of Lima, Peru. Only two houses were not buried or swept away by the landslides.

During the second week of April 1999, mudslides buried several blocks of the town of Argelia, 150 miles west of Bogota, Colombia, resulting in forty-two deaths. Many local residents said the disaster could have been prevented had governmental authorities heeded their warnings that funds were needed to relocate townspeople living in neighborhoods at risk of mudslides.

A week of steady, heavy rains in October 1999 brought on a memorable catastrophe to hundreds of Indians living in a number of remote hillside villages and hamlets 100 miles northeast of Mexico City. In a worst-ever mudslide in that mountainous stretch of Mexico, more than 450 people perished after being buried alive. Another 253,000 dwellers, of which 200,000 became homeless, were affected by the floods. As many as 70,000 of the Indian population had to be housed in shelters. Food, water, and medicine, for the most part, were scarce due to the weather, terrain, and washed out roads. A major concern was the possible breakout of cases of dengue, a mosquito-borne illness, and the spread of skin and respiratory infections.

As often happens after casualties mount in the aftermath of a severe weather-induced tragedy, human blame takes place. In the great Mexican mudslide, officials, media, many who survived, and others faulted the delayed timing of the opening of the dam flows, the lack of an earlier warning system, the prevention of deforestation, and the lack of a governmental program to evacuate homes on the dangerous hillsides to prevent the dreadful event from happening.

Three days of very heavy rainfall, beginning December 15, 1999, set off massive mudslides flooding Venezuela's northern Caribbean coast, causing a costly devastation to many of the ramshackle towns situated on the steep mountainsides ringing Caracas, the nation's largest city. It was the worst natural disaster in that country during the last 100 years.

The nation's authorities estimated that more than 150,000 people were made homeless as a result of the enormous floods and mudslides. Venezuela's grieving government officials predicted that the final death count would surpass 10,000. It was assumed thousands more would never be accounted for. It would take many years to rebuild the destroyed towns for those Venezuelans electing to remain in the area. The government, on the other hand, encouraged thousands of victims to relocate to safer, less urban parts of the country.

Once the rains and mudslides subsided, Venezuelan soldiers and rescue and relief workers in the region encountered uncanny scenes of death and destruction. Cars, buses, tree trunks, boulders, household goods, furniture, and appliances were embedded in twenty feet of mud. Crudely made shacks having sheet metal roofs were driven in total or found in parts a distance from their foundations. Corpses, many dismembered, were scattered all about. The smell of death was dreadful, even penetrating the sanitary masks worn by exhausted volunteers and aid workers.

The people of Venezuela were somewhat consoled by the fact that many compassionate individuals, nations, and relief organizations immediately rushed aid to the victims. Badly needed assistance included financial support, first aid and medical personnel, and vital supplies such as medicine, food, water, clothing, and many kinds of transport vehicles.

WHERE MUDSLIDES MOST OFTEN OCCUR

There is a long history of mudslides that have buried people and homes throughout the world. The villages of the Andes Mountains of South America have been especially hard hit by frequent onslaughts of mud slamming into villages. In the late 1990s hundreds of villagers in Peru were drowned in seas of mud in separate incidents. Mudslides and landslides are common in the Andes during the January to March rainy season.

A landslide is somewhat different from a mudslide. It is a slow or rapid downward movement of rock or soil material on slopes under the influence of gravity. It is usually lubricated by rain water or melting ice or snow. A landslide involves relatively dry masses of earth fragments that rotate backward downhill as they move. They can consist of freefalling debris from cliffs or overhanging slopes; a rockslide of falling or sliding rock masses down bedding, joint, or fault surfaces; or a rockfall—free-falling rocks over a steep slope.

The monsoon season in India usually causes thousands of people to be trapped under homes flattened by mudslides. In the western United States, mudslides and washouts following storms frequently send homes sliding down soggy hillsides and prevent passage on busy roads and highways. Landslides there also block railroad tracks and disrupt passenger service. In Southern California multimillion-dollar estates and expensive homes, situated on steep-sloped foothills, have been turned into rubble by rampaging mudslides.

COPING WITH MUDSLIDES

The victims of mudslides frequently need months for cleanup operations to be completed. Unfortunately for some victims who choose to remain in their same locations, Mother Nature might revisit the site and repeat the devastation. But, more than likely after experiencing two or three mudslide episodes, they too "strike-out" for a more level homesite.

When mudslides bring damages to property, there are on occasions disputes between the insured and the insurance carrier as to the clarity of the insurance coverage. For this reason, one of America's most important insurance companies has recently stipulated the following definition in their policies:

Mudslides (that is, mudflows) that are proximately caused by flooding and are akin to a river of liquid and flowing mud on the surface of normally dry land areas, including your premises, as when earth is carried by a current of water and deposited along the path of the current.

Rescue efforts often prove to be nearly insurmountable tasks owing to terrible road conditions and distances from built-up areas. To compound the problem there is always the fear of outbreaks of diseases and epidemics spreading among the survivors of a stricken village.

Coping with mudslides can often turn into a prolonged and arduous task. When deaths, injuries, and damages are deemed extensive, a state of emergency may be declared by governmental authorities. Funds are usually needed to rebuild an area's infrastructure, provide urgent medical services, and initiate proper sanitary conditions. Institutions, homes, and business have to be rebuilt and reestablished.

People who elect to remain in precarious mudslide zones have to be motivated or required to relocate. This often means financial incentives to the victims or potential targets of mud or rockslides. Even in locations where retaining walls are built to restrain such slides, stress and anxieties persist. When it comes to prevention and prediction of mudslides, the inevitable seems to prevail, particularly where roads and dwellings are situated at the foothills of steep slopes.

MUDSLIDES

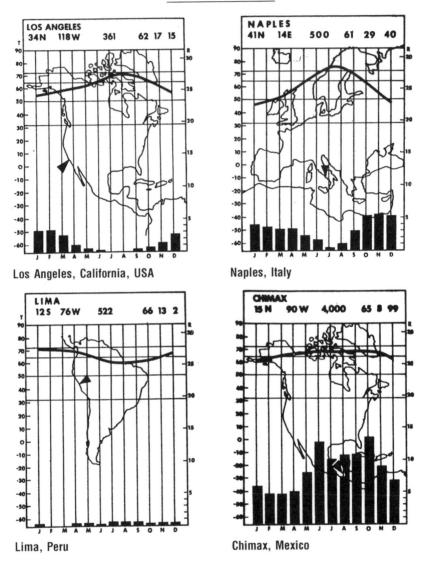

Los Angeles, California, USA

Naples, Italy

Lima, Peru

Chimax, Mexico

People living adjacent to hillsides, as they do in the areas depicted in these climographs (see key in the Introduction), have always been subject to ruinous mudslides during periods of rain and flooding. Mudslides have caused great misery for their victims. They are usually sudden and can be as forceful as a giant bulldozer. Mudslides are indiscriminate in selecting their destructive downward paths. Expensive homes along the California coastal ranges, as well as the flimsy dwellings of the poor mountain dwellers of the Andes Mountains of South America, have been victimized by mudslides.

SOURCES

Astor, Michael. "Brazil Battling Storms, Mudslides." *Sun-Sentinel* (Ft. Lauderdale, Fla.). January 7, 1997.

Barry, R. G. *Mountain Weather and Climate*. London: Methuen, 1981.

Deane, Daniela. "Bodies Abandoned in Mass Mud Grave." *USA Today*. November 4, 1998.

"Death, Devastation, Chaos Leave Venezuela Reeling." *Sun-Sentinel Wire Services* (Ft. Lauderdale, Fla.). December 21, 1999.

Guggenheim, Ken. "Storm's Survivors Stranded and Starving." *USA Today*. November 5, 1998.

Haller, Vera. "South Italian Village Digs Out; At Least 55 Killed in Mudslides After Torrential Rains." *Washington Post*. May 11, 1998.

"Hundreds Missing as Mud Buries Two Villages in Peru." *New York Times*. February 21, 1997.

"Ice and Snow Give Way to Mudslides and Floods." *Sun-Sentinel* (Ft. Lauderdale, Fla.). December 31, 1996.

Kerr, Jennifer. "Storm Sends Five Houses Sliding Down Soggy California Hillslides." *Sun-Sentinel* (Ft. Lauderdale, Fla.). February 8, 1998.

Koop, David. "El Niño Lays City to Waste." *Sun-Sentinel* (Ft. Lauderdale, Fla.). February 7, 1998.

Kovaleski, Serge. "Storm Deaths Put at 7,000." *Washington Post*. November 3, 1998.

"Mudslide Burries Hamlet in Amazon; 15 Dead." *New York Times*. February 15, 1999.

"Mudslides Kills 40 in Bolivia." *Sun-Sentinel* (Ft. Lauderale, Fla.). December 13, 1998.

Natural Disasters of North America (supplement). *National Geographic*. July 1998.

Rohter, Larry. "Fear Erases Hope as Venezuelans Search for Dead." *New York Times*. December 24, 1999.

Schneider, Allen A., ed. *A Dictionary of Basic Geography*. Boston: Allyn and Bacon, 1970.

Stanley, Alessandra. "Italian Town Buries 90 After Mudslide." *New York Times*. May 11, 1998.

Stevenson, Mark. "Mudslide Covers Indian Village." *Sun-Sentinel* (Ft. Lauderdale, Fla.). October 12, 1999.

Rain frequently delays people from getting to and from their workplaces. When the rain is anticipated pedestrians usually prepare for it by dressing properly and by carrying umbrellas. When the rainfall occurs unexpectedly people are often subject to discomfort from the need to get about in wet clothing. (Courtesy of National Oceanic and Atmospheric Administration.)

14

RAINFALL

The poet Henry Wadsworth Longfellow once expressed the idea that "into each life some rain must fall." For some, this excerpt from "The Rainy Day" might imply that rainfall, the water falling to the earth in drops from darkened clouds, may have a negative value. For others a rainfall might be interpreted with a good feeling that the rain means a renewal of life will follow, for without rain crops will not grow, and livestock will be unable to be sustained. Rain, therefore, is a necessity for all life.

Rainfall really means precipitation—generally all the water that falls or forms on an area, including rain, snow (it takes ten or twelve inches of snow to equal one inch of rain), sleet, hail, dew, and frost from the moisture in the atmosphere.

The ascent of moist air, which is the main cause of the formation of clouds, is the sole cause of important amounts of precipitation. After the formation of clouds, continued rapid cooling produces precipitation as it relates to the humidity and cooling of the rising air.

The precipitation may be deemed *convectional rain*, which results when the upward movement of warmer air engages the surrounding cooler air aloft; *orographic rain*, produced when rising ground deflects or uplifts wind-driven moisture laden air; or *cyclonic rain*, caused by the encounters of air currents of different temperatures, such as occurs in the cyclonic storms of varied air mass movements in the middle latitudes. The lines of disparities between the air masses are known as fronts on a weather map.

HIGH AND LOW RAIN AREAS

Rainfall totals are measured in inches or millimeters (.03937 inch). Rain is always falling somewhere on the earth, but there is a huge difference in the amounts that people of certain geographic places experience. On

Mt. Waialeale, on the island of Kauai in Hawaii, the yearly rainfall averages 460 inches. It is probably the wettest place in the world. One year 600 inches fell there. It rains there 355 days of the year. Another extremely wet location is Cherrapunji, India, where the annual rainfall measures 425 inches. During the months of June and July, as much as 203 inches were once recorded. There is also an incredible report that 1,041 inches of rain fell there one time. For Cairo, Egypt, in the north African desert region only 1.1 inches of rain falls. From the three-month period, June to September, no rain is ever recorded. Hardly a trace is ever listed for Iquique, Chile, where a mere 0.05 inches of rain can be measured.

Washington, D.C., the capital of the United States, receives forty-two inches of rainfall annually. Like other cities much of what the people of that area do and how they live on a daily basis can be governed by the amount of rainfall and the times it occurs there. People in any U.S. town find that their social activities and physical involvements may need to be changed by the arrival of unanticipated elements, such as the rain factor.

EFFECTS OF RAINFALL ON DAILY LIVING

Rainfall, or the lack of it, can be the underlying reason people alter their routines, select what to wear, and plan alternate choices in contending with the weather of the hour, day, or week. A barbecue, picnic, family outing, or day at the beach may have to be postponed. Inclement weather might cause the cancellation of their child's soccer or baseball game. If a person walks to his or her workplace or transportation the commuter may need to don rainwear or take along an umbrella. A heavy rainfall can be a factor in deferring the watering of the lawn or garden for another time. It can make car washing an exercise in futility, or help make a decision to leave the marketing or shopping for another day. A sudden downpour may cause neighborhood moms and dads to scoot to the school to pick up the children.

DANGERS OF LARGE RAINFALL AMOUNTS

Falling rain can be fascinating to hear and watch. Rain driving against a window pane or upon a rooftop can have a staccato drumbeat effect that quickly alerts the listener to the dynamic conditions outside the home. Drizzle is hardly noticeable and is rather harmless. Heavy downpours of rain, of the other hand, can be very visible and even dangerous. A good example of this occurred on May 12, 1972, near New Braunfels, Texas, when a single cloudburst dumped sixteen inches of rain on the town in a very short period, sending a thirty-foot wall of rushing water down the local river channel, washing away scores of people, houses, and automobiles. Property damage reached more than $320 million. On that day at that place the rain reigned—unquestionably! Rainfall can readily contribute to wet road problems and automobile accidents due to roads becoming slick and more dangerous, especially right after it starts to rain. On April 15, 1998, a rain-soaked highway (Interstate 70) in O'Fallon, Missouri, caused an eighty-five vehicle pileup, injuring forty people, many seriously. Roof leaks and water seeping into home basements can bring on great concerns and additional expenses. Neighborhood puddles, storm sewer backups, and drainage problems can lead to detours in driving and impede walking.

When rainfall amounts are scarce over a period of weeks a number of serious problems could develop that affect landscape aesthetics, crop yield, and many kinds of economic losses. Drought conditions, especially those that are long-lasting, usually require different, revised practices centered on conservation measures. Drought-stricken gardens and parched lawns often become unproductive or unsightly when water sprinkler systems are curtailed by municipal ordinances. Refillings of swimming pools are discouraged. Sometimes commercial car washing businesses face restrictions. An insufficient rainfall for weeks on end could readily cause wind-blown dust problems in and around the home. Then, there is the possibility of "mud-rains" occurring, which are far from rare and are more commonly found in the central United States. Mud-rain develops when a rainstorm and a dust storm combine. Falling raindrops always pick up some dust particles and when an unusually large amount of dust enters the atmosphere, as in a dust storm, the drops become muddy in color. The actual color depends on the type of soil picked up. They could be gray, yellow, brown, or red, which are often referred to as "blood rains."

RAINFALL

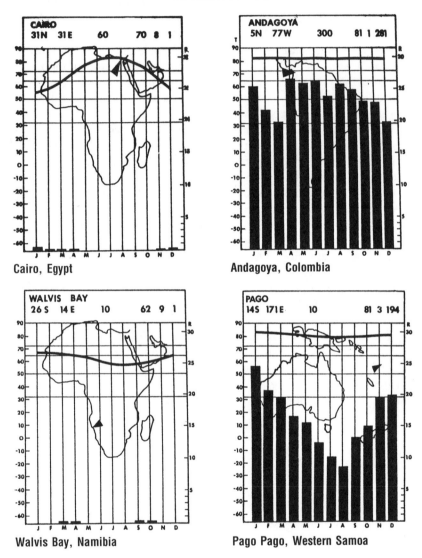

Cairo, Egypt

Andagoya, Colombia

Walvis Bay, Namibia

Pago Pago, Western Samoa

A prime factor in determining the type of lifestyle many people experience the world over is the total amount of rainfall, as well as the time of the year it is received, at their home locations. Note the climographs depicting some geographic places (Cairo and Walvis Bay) that receive as little as an inch annually, and other locations (Andagoya and Pago Pago) that may record hundreds of inches of rainfall throughout the year (see key in the Introduction). The amount of rainfall received is a major factor in determining crop yielding and the kinds of food people eat.

SOURCES

Battan, Louis J. *Weather in Your Life*. New York: W. H. Freeman, 1985.

Belluck, Pam. "21 Dead or Missing as Storms Rage in East and the Midwest." *New York Times*. June 6, 1998.

Blair, Thomas A. *Climatology; General and Regional*. New York; Prentice-Hall, 1942.

Blanchard, Duncan C. *From Raindrops to Volcanoes*. New York: Anchor Books, Doubleday, 1967.

Critchfield, Howard J. *General Climatology* (3rd ed.). Englewood Cliffs, N.J.: Prentice-Hall, 1983.

Fisher, Robert Moore. *How to Know and Predict the Weather*. New York: New American Library, 1953.

Locke, Michelle. "On Pacific Coast, Residents Live on the Edge." *Sun-Sentinel* (Ft. Lauderdale, Fla.). March 7, 1998.

Mason, B. J. *Clouds, Rain and Rainmaking* (2nd ed.). New York: Cambridge University Press, 1975.

Natural Disasters of North America (supplement). *National Geographic*. July 1998.

"Nearly 400 Missing at Sea in Vietnam Storm." *New York Times*. August 21, 1996.

Oliver, John E., and Rhodes W. Fairbridge, eds. *The Encyclopedia of Climatology*. New York: Van Nostrand Rheinhold, 1987.

Pearce, E. A., and C. G. Smith. *The World Weather Guide*. London: Hutchinson, 1984.

Petterssen, Sverre. *Introduction to Meteorology*. New York: McGraw-Hill, 1941.

"Rains Kill 54 in Brazil." *Sun-Sentinel* (Ft. Lauderdale, Fla.). January 5, 1997.

Schneider, Stephen H. *Encyclopedia of Climate and Weather*. New York: Oxford University Press, 1996.

Statistical Abstract of the United States 1998 (118th ed.). Washington, D.C.: U.S. Bureau of the Census, 1998.

White, C. Langdon, George T. Renner, and Henry J. Warman. *Geography: Factors and Concepts*. New York: Appleton-Century-Crofts, 1968.

The World Almanac and Book of Facts. Mahwah, N.J.: Primedia Reference, 1999.

A snowstorm of this magnitude can bring a major city to a virtual standstill for days on end. Vital services and normal activities may be delayed or postponed. Business hours are shortened or curtailed and schools closed. The sanding of roads, application of de-icing chemicals, and snow removal expenses can cost the city millions of dollars, for which they must budget in advance. (Courtesy of National Oceanic and Atmospheric Administration.)

15

SNOW

Millions of people throughout the world have never seen snow. That is because the closer one lives to the Equator, the less likely the chance of snow forming is. Snow can be found there at lower latitudes, however, it would be most unusual to find unless the most curious of snow seekers were willing to hike to very high elevations. At the Equator the snow line (the lower boundary of a high region where snow never melts) is marked at 17,000 feet—a height where few people live. The snow comes from super-cooled droplets of water in clouds, or when the upper air temperatures are below freezing.

Snow is precipitation in the form of white ice crystals, mainly in the shape of branched hexagonal figures, often clustered into snowflakes. No two snowflakes have ever been found to be exactly alike. Snow appears white mainly due to the reflection of light by the tiny surfaces of crystals. Snow may fall when the supercooled water clouds have been converted into ice clouds. One foot of freshly fallen snow equals about the same water content as one inch of rainfall.

For people, children in particular, who have never witnessed snow falling, the initial experience can be very exhilarating and exciting. The snow encountered may arrive in the form of a flurry or shower, brought on by a gust of wind, and ground accumulation could amount to only a dusting or shallow white coating, or the snow deposits may measure any number of inches in height.

DANGERS OF SNOW

The white beauty of freshly fallen snow is often offset by the negative and sometimes dangerous consequences after its initial appearance. Roads may be closed or become slippery. Trees and powerlines can be downed. The weight of heavy snow can collapse a shed or outbuilding. Heavy snow can immobilize a region and paralyze a city. Commuters become stranded. Supplies do not reach their destinations. The cost of snow removal, damage repairs, and business loss can bring about large economic impacts. Major problems in human services can occur when police, fire, and rescue squad vehicles are unable to reach their calls. Some essential workers in hospitals and offices may be unable to report for duty when public transportation ceases, or when snow-clearing crews hold up traffic for hours. It is particularly troublesome when dirty snow piled on the streets begins to melt and cause flooding and ice to form over the surfaces. This can be the cause of numerous falls and some serious injuries to even the most careful of walkers.

The most violent type of snowstorms, the most damaging of all, are blizzards. The latter is carried along by extremely cold winds bearing great quantities of snow. This kind of severe storm registers wind speeds of 35 miles per hour or greater, with accompanying reduction in visibility to less than one-fourth of a mile for at least three hours duration. When snow is combined with sleet—a mixture of rain with snow or hail—the results can be nothing less than hazardous, if not punishing. On November 24, 1977, at Mineral Wells, a small town in Texas, sleet accumulated two inches thick during an ice and snowstorm, requiring the city to spread eighty tons of sand on the streets to make the icy roads passable.

Despite the fact that blizzards usually come after a short period of warm weather, forecasters are fairly successful in telling when and where they will strike. Humans and animals are prone to losing their lives when they go out into the depth of the blizzard. They can quickly become exhausted by the driving wind, blinded by snow and ice particles, become immobilized and freeze to death. Especially tragic are the casualties of people who become disoriented after they abandon their automobiles on snow-bound highways, or even venture out of their homes to walk to a nearby store. To compound the problems brought on by snow events, there has always been a high incidence of deaths from heart attacks when, unwisely, people try to hand shovel their walkways, driveways, and snowed-in cars. This kind of exertion puts a tremendous strain on the heart muscle.

HISTORIC BLIZZARDS AND SNOWSTORMS

Memorable blizzards have crippled sections of the United States over the years and have paralyzed normal activities for lengthy periods, inflicting death and destruction to areas affected. The most infamous blizzard of all took place over large expanses of the east on March 11–14, 1888. Known as the Blizzard of '88, it caused over 400 deaths and has been used as a kind of standard to compare subsequent blizzards to. Major blizzards since then have claimed thousands of lives and huge property and business losses. Lesser localized blizzards over the years have added countless casualties to this number. Many weather stations in the conterminous United States have reported over 400 inches of snowfall per year. Historically, the greatest annual snowfall recorded was on Mt. Rainier, Washington, in 1955–1956 when 1,000 inches were measured.

Over the years snowstorms seem to have a way of providing some uncanny surprises and freakish outcomes to the people affected. During the great American movement westward a monstrous winter snowstorm in 1846–1847 trapped eighty-seven pioneers enroute to California, in a high mountain pass, now known as Donner Pass, for many weeks. The forty-seven that survived were driven to cannibalism before being rescued. Fortunately those stranded by snowstorms now often resort to more civilized ways of saving themselves. They use snow as an insulator against deep ground frost and cold by building shelters, igloo style. They melt the snow for their water supply and send messages etched in the snow to reveal their locations to airborne rescuers. Now hikers, snowmobilers, campers, and car travelers, marooned by sudden snowstorms, can often rely on modern sophisticated communication devices to be located and saved from dire consequences.

A most unusual snowstorm struck Jerusalem, Israel, during the early hours of January 12, 1998, a desert-bound city where the minimum average daily temperature never falls below 41 degrees F, well above freezing. But on that day more than a foot of snow fell. For most Jerusalem residents it was their very first glimpse of snow. The rare snow turned the Holy City into one of a carnival atmosphere. Children awoke to an entire whitened landscape that some labeled "cotton land." Schools closed; street traffic was almost entirely halted since there were no snow plows on hand in the entire city to remove the foot of snow that had fallen. Palm trees blanketed with snow swayed in the wind. Friendly snowball fights erupted. Much of the city's normal life was disrupted or shut down. And to join in the festive air, tourists built a snowman at the site of the sacred Western Wall.

Unpredictable snowstorms can catch people and city officials off-guard, causing everyone to scurry about in an effort to contend with it. In February 1887, from four to seven inches of snow fell in the San Francisco Bay area, which was most extraordinary for that city. Since residents had little in the way of clothing or overshoes for snow conditions, the proper garments had to be rushed to the local stores from colder mountainous regions.

Surprise snowstorms have proven to be major nightmares for unsuspecting cities, often bringing them to a complete standstill for days. An early trace of snow fell on Chicago on September 25, both in 1928 and 1942, the earliest snow ever recorded there. In New York City, on December 26, 1947, 30,000 people were put to work to remove twenty-six inches of snow that fell upon the city. Removing the 200 million tons of snow cost more than $8 million. Boston, Massachusetts, has had an unenviable reputation for receiving blizzards, dumping record amounts of snow (twenty-five to thirty-five inches) upon the city as it did in February 1978 and 1983, essentially making the city helpless for almost an entire week. Snow can also arrive late in the winter season as it did in Chicago on March 17, 1991, when fourteen inches of snow—not rain—fell on the St. Patrick's Day parade, causing its postponement. And snow can linger on for snow hikers as it often does on snowclad Mt. Rainier, Washington, until the first week in August.

EFFECTS OF SNOW ON DAILY LIVING

The vision of snow as an asset or liability is in the eyes of the beholder. Children, but hardly their parents, tend to be overjoyed when a snowfall closes the schools. They can romp in the snow, build snowmen and forts, sleigh ride, and of course, engage in mischievous snowball throwing. Older children can readily hire themselves out as snow shovelers and earn a considerable amount of money. For certain businesses a snowstorm can increase their sales of snowmobiles, snowblowers, shovels, and chemical salts. Clothing stores stocking special snow apparel can do well. On the other hand, many stores, offices, restaurants, and gas stations find their business suffers when a snowstorm tends to confine people to their homes until conditions return to normal. Even ski resorts lose out on income revenues when a snowstorm blankets a region, preventing skiers from reaching the slopes by air or surface means. This is what happened to northern California ski centers when ninety-six inches of snow fell on the region, along with 90 mile per hour winds, and closed the mountain resorts in early March 1967. The outlay of costs for snow removal by private and governmental agencies can be enormous.

Americans, perhaps fascinated by snow, have included a number of snow-based terms in their everyday vocabulary. Words and terms such as snowball (to increase rapidly), snowbird (people who migrate to warmer climates during the winter season), snow job (the act of deceiving by glib talk), snow white (white as snow) and snow-blind (blinded temporarily by snow reflections) are in common usage today.

So far there is no snow term coined that is applicable to describe the artificially induced snowfalls made by sprinkling particles of dry ice seedings into clouds in order to form frozen crystalline flakes. This is what was tried with mixed success in nine northern provinces across the steppes of Mongolia to put out huge fires that raged there in early May of 1996. The fires blazed for more than four weeks and killed seventeen people and forced 2,200 nomad families and more than one million cattle to be evacuated before the man-made snow helped put out the burning forests and pastureland.

PREDICTING AND PREPARING FOR SNOWFALL

For the amateur weather forecaster concerned about a possible snowfall there are a number of visual guidelines to look for. A home barometer would enhance the layman's prediction. The temperature of air must be at or below 32 degrees F. A rapidly falling barometer, usually decreasing below a sea-level pressure of 29.92 inches, or 1013.2 millibars, and winds out of the north or east (especially east of the Rocky Mountains in the United States), along with thick, lowering clouds, or ragged-based wet-looking dark nimbostratus are indicative, if not ominous, signs of an impending snowstorm. A sudden drop in temperature, with an increase of wind and an encroaching cold front from the north or west would give added credence to the prediction.

When these time-proven weather signs appear preparations should be made to be ready with the snowblower, snow shovel and salt-melting pellets. It may also be the time for the skis, sled, parkas, boots, and snowsuits to come out of hibernation.

SNOW

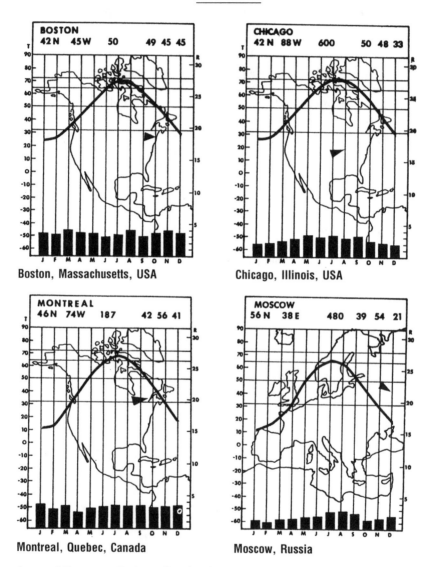

Boston, Massachusetts, USA

Chicago, Illinois, USA

Montreal, Quebec, Canada

Moscow, Russia

A snowfall can readily beautify a landscape in the countryside. A snowstorm in a city can be a major liability and may result in the impediment of normal municipal services and essential operations. For the most part, children enjoy the exhilaration of a snowstorm. Adults may have a different perception of the merits of snow. Snow, therefore, is in the eyes of the beholder. Snowfalls are a normal occurrence during the winter months in and around the cities identified by these climographs, which show substantial precipitation in months where the temperature curve dips below freezing (see key in the Introduction).

SOURCES

"Andes Snowstorm Traps Thousands in Peru." *New York Times*. August 12, 1997.

"Big Snowstorm Cows Buffalo." *Sun-Sentinel* (Ft. Lauderdale, Fla.). January 2, 1997.

Egan, Timothy. "In Spokane, Snow and Ice, Then Falling Trees." *New York Times*. November 27, 1996.

Fisher, Robert Moore. *How to Know and Predict the Weather*. New York: New American Library, 1953.

Graf, Daniel, William Gartner, and Paul Kocin. "Snow." *Weatherwise*. December 1995/January 1996.

Kahl, Jonathan D. *Wet Weather: Rain Showers and Snowfall*. Minneapolis: Lerner Publications, 1993.

Kifner, John, "The Roar of Cold Cash on Snowy Maine Trails." *New York Times*. February 1, 1997.

Ludlum, David M. *The American Weather Book*. Boston: Houghton Mifflin, 1982.

McIllneen, R. *Basic Meteorology*. New York: Van Nostrand Reinhold, 1986.

National Oceanic and Atmospheric Administration. *Are You Ready for a Winter Storm?* Washington, D.C.: NOAA, 1991.

Natural Disasters of North America (supplement). *National Geographic*. July 1998.

Petterssen, Sverre. *Introduction to Meteorology*. New York: McGraw-Hill, 1941.

Rimer, Sara. "April Brings Cold Surprise to Blanket New England." *New York Times*. April 2, 1997.

————. "Snowplows Finally Clear Boston, but the Anger Lingers On." *New York Times*. April 5, 1997.

Schneider, Stephen H., ed. *Encyclopedia of Climate and Weather*. New York: Oxford University Press, 1994.

"Snow Blankets Jerusalem, Middle East." *New York Times*. January 13, 1998.

"Snow Made to Halt Fire Kills Cattle in Mongolia." *New York Times*. May 12, 1996.

"Winter Makes Early Debut In The Midwest." *New York Times*. November 12, 1996.

"Winter Storms—the Deceptive Killers." A Guide to Survival. National Oceanic and Atmospheric Administration, National Weather Service, November 1991.

Sunshine can have its effects even in cold weather. Here cross-country skiers enjoy the bright sunshine that reflects on the snow. (*Credit:* AP/Wide World Photos. Reprinted with permission.)

16

SUNSHINE

The chances are almost certain that residents waking up in the Tucson and Yuma, Arizona, areas in the American southwest desert, upon turning on their television or radio for a weather report, will hear the forecaster declare that "today will be another sunny day." This area receives sunshine on average 94 percent of the possible days during the year. Nearby Phoenix has an annual 85 percent sunshine possibility. But for those living in Vancouver, Canada, or Edinburgh, United Kingdom, the percentage of possible sunny days is only 31 and 33 respectively on a yearly average. For people who choose to live in a place where there is a moderate amount of daily sunshine, cities like St. Louis, Missouri, or Little Rock, Arkansas, may be more to their liking. There the sun tends to shine about two-thirds of the time.

The sun's rays are directly over some point on the earth between the Tropic of Cancer at 23.5 degrees north latitude on or about June 21 and the Tropic of Capricorn at 23.5 degrees south latitude on or about December 21. The sun is always directly over the Equator twice each year, once on or about March 21 and once on or about September 21. These changes of the position of the earth in relationship to the sun cause the seasons. The seasons, accordingly, are most instrumental in determining overall temperatures and thus a major factor contributing to man's activities and the ways he lives in both the Northern and Southern Hemispheres.

DANGERS OF THE SUN

Many people seem to crave a "place in the sun." The old adage that there is "nothing new under the sun" is now being proven to be wrong in one respect. Medical research experts have found that there is a serious risk for prolonged exposure to the sun, including an increased incidence of skin cancer and cataracts, an eye disease. For this reason it is a routine practice for people who spend sun-filled days at the swimming pool or at the beach to try to protect themselves from the ultraviolet rays of the sun by applying suntan lotions and sunscreens to their bodies. Research studies of late have found that skin exposure at low latitudes, high altitudes, and at beaches having light-colored sand with adjacent water reflections increases the likelihood of getting a painful sunburn or absorbing harmful ultraviolet radiation into the body.

The sun's rays can also prove to be a threat to the health and normal eye function due to glare. For those driving a car into bright sunlight, or for those sun-loving activists "out in the sun" for long periods of work or play, sunglasses help to shield the sun's glare for more efficient vision.

EFFECTS OF THE SUN IN DAILY LIVING

The sun is easily the most important of all the celestial bodies to us on earth. It makes our weather. "All the world is cheered by the sun." Its heat and light are absolutely essential for the life and health of man, animals, and plants. It determines our weather, and its rays store energy in our food like vitamin D, the "sunshine vitamin." We use the sun's energy in a number of ways that are not usually apparent. Coal and petroleum, essential for power and fuel, emanate from sun-drenched plants embedded in the earth millions of years ago. Since the sun makes the winds, windmills can be a source of power for grinding grain and pumping water.

The sun is responsible for lifting water into the atmosphere and converting the moisture into rain, which in turn eventually finds its way into water bodies and can be retained by dams prior to turning turbines to generate electricity. Hardly anyone realizes that when we sit by a campfire or home fireplace the flames that glow and the warmth sent forth is a release of light and heat once stored by a tree when it was alive. It would, therefore, be extremely challenging to try to find anything—animal, vegetable, or mineral—that would be unaffected by sunlight.

Amid concerns about future power needs, the geometric growth of the world's population and the limits and dangers of reliance on petroleum, coal, and nuclear energy and the potential for developing ways to harvest sun or solar power are taking on new research priorities. In 1952 Israel began to install the world's first rooftop solar collector panels to capture the sun's valuable rays for domestic use. Today in Israel the panels are everywhere, providing hot water to 70 percent of the country's population. Even without Israel's high rate of sunshine, Switzerland uses many solar energy stations to light up the country's highways and billboards. Similar systems now operate in parts of Italy, Spain, and Japan. California, a state with sunny deserts like Israel, leads the world in solar-generated electricity.

SUN IN MYTH AND FOLKLORE

The sun, from primitive life to the present, as it changed between day and night and as a provider of warmth, or lack of it at certain times, has always been the object of mystery, superstition, mythical explanations, and frequently as a god to be worshiped. Ancient and even more recent cultures often paid homage to it through ritualistic religious dances. In North America the Plains Indians performed tribal sun dances, some of which were savage, in honor of their sun god, hoping their brightly colored painted bodies would draw the god's attention and bring them luck.

Today, even sophisticated and well-informed people in many diverse societies participate in ceremonies to acknowledge a special day that recognizes a change in season or sun's "position" in the sky, as the earth makes its annual orbit around the sun. In the northern most villages of Norway, well beyond the Arctic circle, it is traditional for local villagers to turn out on Sun Day to greet the return of the sun. On January 21, 1998, the inhabitants of Tromso, by now tired of weeks of almost total darkness, assembled at the lake speculating on the arrival time of the sun. At twelve minutes before noon the sun popped out, like a streak of fire spread across the sky, visible for a fleeting twenty minutes. No small wonder why all assembled let out a cheer.

The sun, visible at midnight in the Arctic or sub-Arctic regions during the summer, is described as the land of the "midnight sun." To offset that, the converse season brings extended periods of no or little sunlight in the winter months, depending on similar latitudinal locations. For example, the possible duration of sunshine at two northern latitudes are the longest day and the shortest day (of light) of the year: at 10 degrees north of the Equator there are 12 hours 35 minutes of summer daylight, but at 90 degrees on the same day usually around June 21 the sunshine lasts for 24 hours. At the same latitudes the shortest day's duration of sunlight is for 11 hours 25 minutes while at 90 degrees there is no or little sunshine hours during the winter. The reason for this is that the earth is inclined toward or away from the sun at a tilt of 23.5 degrees in its revolution around the sun. That is why there is a wide difference in the amount of sunlight hours people have come to rely on during the four seasons of the year. The sun plays a major role in their lives, depending on their latitudinal address. It can have a bearing on how people adjust their daily activities depending on their exposure to cold, heat, or extended periods of daylight and darkness.

A CYCLICAL MULTIPURPOSE SUN

Man has learned to rely on the sun as a kind of "calendar reminder" to make provisions for the future if he is to survive. As a climatic element the sun provides us with a seasonal rhythm for which man has had to adjust. Besides the effects of the sun in our daily living we knows that "there is a time to sow and a time to reap." We realizes that the sun's impact is cyclical and we must be ready for it at all times.

The sun can be a weather predictor. The "feel" of the sun's temperature and the sighting of it angle above the horizon can serve as indicators of the coming seasons. It can easily serve as a daily weather forecaster. When the setting sun looks like a ball of fire and the sky is clear, and the wind velocity light, the next day will bring fair weather.

The sun has entered the realm of art as well. Since ancient times artists have depicted the sun in pictures, carvings, sculptures, and metal work. Artists, like Vincent Van Gogh have attempted to capture the beauty, color, and warmth of the sun on their canvases for posterity to enjoy. Music too has been enriched as a result of man's admiration of the sun by the countless songs written by famous composers, like Nikolay Rimski-Korsakov.

SUNSHINE

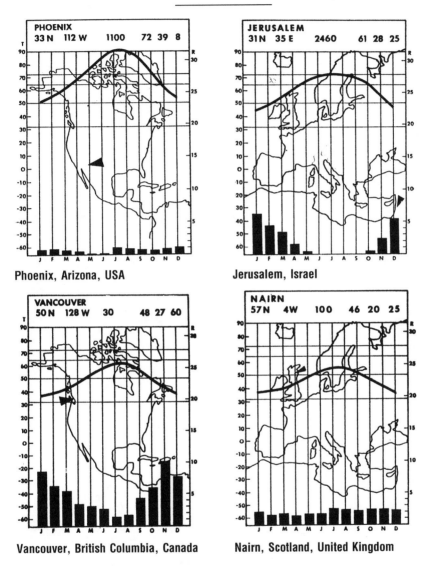

Phoenix, Arizona, USA

Jerusalem, Israel

Vancouver, British Columbia, Canada

Nairn, Scotland, United Kingdom

There is a wide difference in the amount of sunlight hours people have come to rely on during the four seasons of the year, as can be noted in these climographs (see key in the Introduction). The sun plays a major role in our lives. It can have a bearing on how people adjust their daily activities depending on their exposure to cold, heat, or extended periods of daylight or darkness. Despite the fact that people often choose to follow the warmth of the sun to escape winter's cold, it is now realized that too much sun can be detrimental to one's skin and cause serious health problems.

SOURCES

Blair, Thomas A. *Climatology; General and Regional*. New York: Prentice-Hall, 1942.

Fisher, Robert Moore. *How to Know and Predict the Weather*. New York: New American Library, 1953.

Fishkoff, Sue. "Turn on the Sun." *Jerusalem Post International Division*. February 1, 1997.

Funkle, Phyllis Ellen. "Night for Day." *Sun-Sentinel* (Ft. Lauderdale, Fla.). February 23, 1997.

Giese, Arthur C. *Living with Our Sun's Ultraviolet Rays*. New York: Plenum Press, 1976.

Halacy, D. S., Jr. *Earth, Water, Wind and Sun: Our Energy Alternatives*. New York: Harper and Row, 1977.

Herman, John R., and Richard A. Goldberg. *Sun, Weather, and Climate*. Mineola, N.Y.: Dover, 1985.

Meinel, Aden, and Marjorie Meinel. *Sunsets, Twilights and Evening Skies*. New York: Cambridge University Press, 1983.

Naseri, Muthena, and Douglas Smith. "Solar Energy," in *Environmental Encyclopedia*. Detroit: Gale Research, 1994.

National Disasters of North America (supplement). *National Geographic*. July 1998.

Persinger, M. A. *The Weather Matrix and Human Behavior*. New York: Praeger, 1980.

Schneider, Stephen H., ed. *Encyclopedia of Climate and Weather*. New York: Oxford University Press, 1994.

Statistical Abstract of the United States (118th ed.). Washington, D.C.: U.S. Bureau of the Census, 1998.

"Sun, Wind, Powering the Future, Says Group." *Home News and Tribune*. (New Brunswick, N.J.). October 16, 1994.

17

TEMPERATURES (COLD)

A "cold temperature" is a very subjective term. It could mean the temperature felt by the human body, called the sensible temperature. It could infer the free air temperature of the atmosphere. Cold temperatures are in the main caused by one or more of the following climatic controls: (1) latitude, (2) site and situation, (3) land or water proximity, (4) seasonal and diurnal changes in relationship to the sun, (5) pressure belts, (6) air mass characteristics and movements, (7) ocean currents, (8) topography, (9) wind direction and velocity, (10) elevation, (11) humidity, or (12) radiation of sun and earth.

For people living in cold climates or for those experiencing a winter cold wave, there is frequently an attempt made to describe how cold they feel. Such expressions as "deep freeze," "numb with cold," "bone chilling," or "shivering cold" are cliches that are used to reinforce their comparisons, sometimes as a benevolent warning to friends and relatives to stay indoors during the subfreezing weather. The well-meaning intentions would probably go unheeded by people in places like Alaska, Canada, Scandinavia, or most certainly Siberia ("the refrigerator of the world"), where very severe cold temperatures last for months on end and are a condition they have become acclimated to.

Some of the lowest known temperatures in the world occur in Verkhoyansk in the Yakut region of Siberia. The average there is −58 degrees F in January, but an occasional "high" reading of −90 degrees F (in this case high means low) during that month would fail to startle the local weather observer. Oymyakon, Siberia, holds the record for being the coldest inhabited place in the world. The temperature reached −96 degrees F there in 1964. A temperature of −128 degrees F was recorded at research station Vostok, Antarctica, on July 21, 1983, a place too cold for human settlement. It is quite obvious that no matter where you live in Asia, north of 55 degrees latitude, it is impossible to escape the snow and cold of the infamous Siberian or Russian winter.

Extreme cold often accompanies a winter storm or is left in its wake. Prolonged exposure to the cold can cause frostbite or hypothermia and become life-threatening. Infants and elderly people are most susceptible. What constitutes extreme cold and its effect varies across different areas of the United States. In areas unaccustomed to winter weather, near freezing temperatures are considered "extreme cold." Freezing temperatures can cause severe damage to citrus fruit crops and other vegetation. Pipes may freeze and burst in homes that are poorly insulated or without heat. In the north, below zero temperatures may be considered as "extreme cold." Long cold spells can cause rivers to freeze, disrupting shipping. Ice jams may form and lead to flooding. (Courtesy of National Oceanic and Atmospheric Administration.)

On an average winter day in Yakutsk, for example, when the temperature is −46 degrees F, the air singes the skin. Walking is fast, but talking is slow. A Yakut needs to catch his breath in order to do either. Where mouths are uncovered, mist is dispelled into the biting heavy air. The clothing people wear is heavy too. Children are wrapped in layers, sweaters, and outer garments, with their scarves holding the coat collars close to the face to keep the flesh from receiving freeze burns. Two sets of gloves are worn, one wool and the other fur. Adults wear boots, leg wraps, thick wool socks, and flannel undergarments. Fur-lined pants, wool shirts and sweaters, and hooded parkas of leather cover the entire attire. Men don head coverings made of local animal furs, and women rely on the multipurpose kerchief—or babushka—tied under the chin to keep their heads warm.

It seems that Siberians, like all people exposed to severe cold, know the importance of keeping their heads covered since 80 percent of all body heat escapes from the top of the head. What they fail to realize, however, is that they need to limit their consumption of alcohol, which robs the body of vital heat. In regard to food consumption, persons living in Arctic-like climates require a greater intake of calories than warmer climatic regions. Generally, caloric needs can increase by 2,000 during an extended stay in cold weather.

A century ago, when criminals and political prisoners were marched off into the wilds of the frigid east and exiled to "the land of the dead," as Siberia was called, it was often necessary for them to hibernate for the entire winter in some remote makeshift hut in the woods in order to stay alive. Still, hundreds of thousands perished as their shelters proved no match against the cold. Now, life expectancy in Siberia tends to be greater than most regions of Russia. In 1995 Russian life expectancy at birth for males was sixty-four years and seventy-four years for females, relatively low for an industrialized, advanced country. In the United States men were expected to live until age seventy-three and women age eighty. Siberia boasts of having the world's highest percentage of people living beyond one hundred years of age. It is claimed that one Yakutsk native allegedly lived through 154 winters. One theory advanced for Siberia's long life rate is that in such a frigid environment it is almost impossible for bacteria to affect vegetable and animal life.

Most people of the region spend more than 85 percent of their waking hours indoors from November to February. For those who need to be outdoors for long periods of time, warm-up shacks are made available for ten minutes every hour, otherwise, the human body may become deprived of feeling and incapable of motion. That is why it is not rare in Siberia to see a disproportionate number of people who have suffered limb amputations due to previous encounters with frostbite. Frostbite is tissue damage to a part of the body that has been exposed for too long to intense cold.

Hardy, cold-resistant Siberians have learned to function in such penetrating cold. Even the children are required to attend school under these severe condition, unless the temperature falls below −45 degrees F, when schools close. Perhaps that is why Siberian youngsters often tend to be able to read a thermometer at an early age.

It would be quite misleading to conclude that record-setting cold snaps are exclusive to the domain of Siberia. Cold weather has been blamed for many deaths in such unlikely countries as Mexico, where poor people living in mud-and-thatch huts or in shacks lacking adequate heating were particularly vulnerable. The European continent has suffered numerous deadly cold spells over the years. In January 1997 more than 230 people died there as the result of a cold air mass that lasted about two weeks. Many that perished included the elderly, poor, and homeless. With temperatures at −14 degrees F, as it was in Paris during this deep freeze, the human heart is under a great strain to perform as well as it would in warmer, more normal, free-air temperatures. People exposed to extreme cold are at a greater risk of having a heart attack.

For the less fortunate that have to sleep out in the open, the frigid air is definitely more perilous when there is a wind-chill factor. Both temperature and wind cause heat loss from body surfaces. A combination of cold and wind makes a body feel colder than the actual temperature. This is known as the wind-chill factor. A temperature of 20 degrees F, for example, plus a wind of 20 miles per hour, causes a body heat loss equal to that in −10 degrees F with no wind. Essentially the 20 degrees feel like −10 F. Under these dangerous situations, the police and social workers of cities that have homeless people sleeping out in the streets and parks endeavor to round them up and take them to indoor shelters for the night. People exposed to a prolonged period of extreme cold, especially the elderly or infants, can readily get a life-threatening condition called hypothermia, which is a subnormal body temperature.

Cold fronts can play tricks on people and places depending upon the movement of the air mass and the local topography. On some occasions the freakish weather disparity between two nearby towns cannot be explained. On January 20, 1991, the temperature from the towns of Lead and Deadwood, South Dakota, only two miles apart, recorded a wide difference in temperature readings taken at the same time—52 degrees and −16 degrees F. South Dakotans traveling between the two towns were confronted with an understandable case of sudden "weather shock," necessitating a quick change in the clothing they were wearing at the time and much bewilderment, to say the least.

Being left out in the cold, spending two days hiding in freezing temperatures, led to a lifesaving decision by a jail escapee very anxious to turn himself in and return to the warmth of his jail cell. On January 19, 1997, near Valdosta, Georgia, a twenty-year-old prisoner who had fled a jail there kept knocking on the doors of frightened residents pleading for them to call the police so that he would be taken back into custody and a warm jail cell, which is what happened after a 911 call was made. Since the prisoner learned some cold facts about escaping during bitter cold nights, it might be advisable for him to make wiser long-term plans. Should he ever be released from prison and decide to rehabilitate himself, he might want to choose to live in a more temperature friendly community like Winter Haven, Florida, or Sun City, Arizona.

TEMPERATURES (COLD)

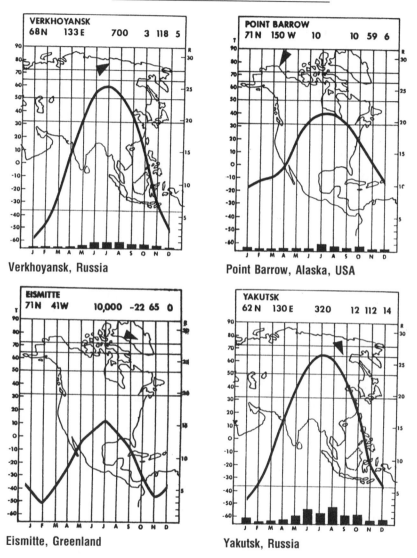

Verkhoyansk, Russia

Point Barrow, Alaska, USA

Eismitte, Greenland

Yakutsk, Russia

For people living in severe cold climates where very low temperatures last for long periods of time such as those depicted in these climographs (see key in the Introduction), extreme measures need to be taken in order to protect themselves, particularly in terms of clothing selection, shelters, and prolonged exposures to subzero temperatures. Contending with the cold in such places is often a matter of conditioning oneself to basic methods of survival.

SOURCES

"Cold Kills 200 Bangladeshis." *Sun-Sentinel* (Ft. Lauderdale, Fla.). January 29, 1997.

"Death Toll Rises Above 220 in 11th Day of Europe's Freeze." *New York Times*. January 4, 1997.

Farah, Mounir A., et al. *Global Insights*. New York: Glencoe, Macmillan/McGraw-Hill, 1994.

"Freedom Cold, Jail Warm." *Sun-Sentinel* (Ft. Lauderdale, Fla.). January 20, 1997.

Hickcox, David H. "Temperature Extremes." *Weatherwise*. February/March 1996.

Marshall, Steve. "Ice-cold Weather Clamps Down on Midwest." *USA Today*. December 18, 1996.

National Disasters of North America (supplement). *National Geographic*. July 1998.

Paddock, Richard C. "Weather Hampers Search for Victims in Plane Disaster." *Sun-Sentinel* (Ft. Lauderdale, Fla.). December 8, 1997.

Pearce, E. A., and C. G. Smith. *The World Weather Guide*. London: Hutchinson, 1984.

Resnick, Abraham. *Siberia and the Soviet Far East: Endless Frontiers*. Moscow: Novosti Press Agency Publishing House, 1983.

Shacham, Mordeckai. "Danger by the Numbers: Meaningful Cold Weather Indications." *Weatherwise*. October/November, 1995.

Specter, Michael. "Arctic Tribe's Hard Life Unchanged for Centuries." *New York Times*. November 22, 1994.

Stanley, Alessandara. "Moscow Is Very, Very Cold (and a Bit Cold-Hearted)." *New York Times*. December 17, 1997.

Trotta, Dan. "Record-setting Cold Snap Kills 28 in Mexico." *Sun-Sentinel* (Ft. Lauderdale, Fla.). December 17, 1997.

Waters, Steve. "Hypothermia Can Be a Killer." *Sun-Sentinel* (Ft. Lauderdale, Fla.). January 4, 1996.

White, C. Langdon, George T. Renner, and Henry J. Warman. *Geography: Factors and Concepts*. New York: Appleton-Century-Crofts, 1968.

18

TEMPERATURES (HOT)

Generally the same climatic controls that influence cold temperatures also apply to hot temperatures. The sun, however, is the major agent in determining the heat gradient for people and places. Feeling the heat, not unlike feeling the cold, is similarly a state of mind or the temperament of the person affected.

It seems everyone loves to talk about how the weather is changing, or how different it is compared to how they remembered it to be in years past. Some justify their views by citing the signs of climate change recognized by many scientists who contend that a global warming trend is starting to take place. They provide research data that show that the atmosphere has warmed by about one degree in the twentieth century, and that the global sea level has risen by up to ten inches in the past 100 years due to a mean temperature increase. They also attribute the 10 percent rise in the frequency of rainstorms and snowstorms, as well as the upswing in the frequency of winter cyclones in recent years to the worldwide warming phenomenon. In the United States the hottest July ever was recorded in 1998.

The climate change theory is conveniently offered as a rationale when explaining prolonged heat waves that periodically stifle various regions of the country during the summer months. Persistently high temperatures over an extended period coupled with abnormally high humidity reduces the body's ability to cool itself, causing a wide range of adverse impacts on human health, even death. People become greatly enervated when the heat index—what hot weather "feels like" to the average person—reveals high readings for both temperature and humidity. A case in point indicates that the body's sensible temperature would feel like it was 136 degrees F when the real temperature is 95 degrees F and the relative humidity 80 percent. Sunstroke and heat exhaustion are likely when the heat index reaches 105 degrees F.

Sunstroke is caused by excessive exposure to the sun and is charac-

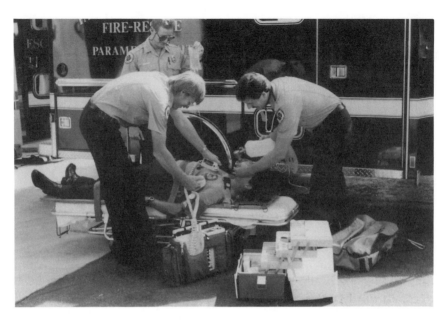

People exposed to prolonged high temperatures can suffer from heat exhaustion, dehydration, heat stroke, and, particularly common among the elderly, heat suffocation. Here a paramedic team demonstrates first aid procedures for heat victims. (Courtesy of the Fire-Rescue Training and Safety Division, West Palm Beach, Florida.)

terized by high body temperature and collapse. Heat exhaustion is a mild form of heatstroke, characterized by faintness, dizziness, and heavy sweating.

Extremely high temperatures can quickly become life threatening, soaring to well above 140 degrees F when free air movement is closed off in a confined area. Pets and children have been known to suffocate and die when left unattended in automobiles with closed windows during excessively high temperatures.

Heat waves extending over a period of days can readily kill people, particularly the elderly or those having respiratory problems. It is the leading weather-related factor in deaths in the United States. It is estimated that some 175 to 1,500 Americans die each year from excessive temperatures. Prior to the advent of air conditioning, the toll was appreciably higher. During the very hot summers of 1930–1936, it is estimated some 15,000 died from heat-related health causes. On July 9, 1936, New York City recorded its hottest temperature ever, a blistering 106 degrees F. The city's second hottest day came on July 21, 1977, when temperatures reached 104 degrees F. Weather experts speculated the temperature to be even considerably higher on the streets of Manhattan where skyscraper buildings tend to block off helpful winds. In 1977 New Yorkers, trying their best to beat the heat, took refuge in air-conditioned offices, walked on the shady sides of streets, dressed in light-colored, loose clothing, drank plenty of liquids, and took it upon themselves to create "do-it-yourself pools" by opening thousands of fire hydrants in order to cool off. The number of heat-related deaths per year in New York averages 320.

Urban areas are hit particularly hard during heat waves because they become "heat islands," where paved streets and brick houses retain heat and there are few trees or bodies of water to provide relief. Heat-related health problems are more severe in cities with many row houses, like Philadelphia, where the wall-to-wall structures with few windows and black tar roofs can readily become "brick ovens." In cities where elderly people and the infirm become "shut-ins," the death tolls from the heat can be either a primary or secondary cause of their demise. Many older persons living in less-advantaged neighborhoods, without air conditioning or fans, shut or bar their windows and bolt their doors in fear of unwanted intruders, and they can easily succumb to the suffocating hot air in their closed apartments.

The city of Chicago suffered a brutal invasion of high heat and humidity over a three-day period in mid-July 1995, in which as many as 733 Chicagoans may have succumbed to the "heat storm," as it was called. The city's entire medical trauma system was overwhelmed as hundreds were sent to hospital emergency rooms. Funeral homes had to order extra coffins and hold services late into the night. Because of the Chicago tragedy, that city and other major U.S. cities began to study and improve ways to be ready to administer to people during heat waves in order to prevent similar mass deaths. Cooling centers have been set up with efficient means of transporting potential victims. City workers are on standby alert to be dispatched to homes to knock on doors of the elderly, deliver food, and provide medical assistance.

Heat waves know no political boundaries. Texas and Oklahoma, along with ten other southern and southwestern states, were declared a disaster area because of a long heat wave and drought in June and July 1998, which caused $1.5 billion in agricultural damages. Some parts of the region had more than forty days of 100 degrees-plus weather. At least 100 people died. It was so hot that railroad tracks warped. In 1980 the Dallas-Fort Worth area had sixty-nine days when the temperature was above 100 degrees F, including forty-two consecutive days. There were seventeen deaths in Dallas alone. The country of India tends to suffer from frequent summer heat waves. Each year, it seems, hundreds die when temperatures soar to 120 degrees F in some places as it did in northeastern India late in May 1998. In Europe, where there are considerably less air conditioned or fan-cooled homes and buildings, a heat wave, though relatively rare, can be very discomforting to those that need to endure the uncomfortable effect of the sweltering heat. That is when everyone who is able heads for the seaside where beaches become obscured by the density of mankind.

When the mercury in a thermometer rises to and above the 90 to 95 degree F range, there can be a wide array of adverse impacts on human and public health. Garbage storage and disposal could become a significant health disorder if neglected. The spread of infectious diseases, like malaria, needs to be controlled, and mosquito-infested swamps have to be sprayed with chemicals. It is also believed that parasitic ticks, which have become a serious problem, increase in number during heat outbreaks.

Too much heat can cause any number of illnesses, some potentially fatal. Spending too much time in the heat can lead to heat cramps, heat exhaustion, heatstroke, and dehydration when the body loses much fluid through sweating. If precautions are taken, one can easily avoid getting sick from the heat. Activities that deplete energy should be done in the cool morning or evening hours. Light-weight fabrics that "breathe" and a hat to keep the head cool and shield your eyes from the sun should be worn. Cool showers should be taken as often as possible, and ice pops, ice cream, and lots of cold juicy fresh fruits should be eaten. Most important of all you should drink lots of water, even if you aren't thirsty. If you do sports activities or exercise in the heat, carry a water container with you and drink up to nine to twelve cups of water for each hour of activity depending on your weight.

People throughout America are more aware of the weather and are more temperature-conscious than every before. Television and radio convey frequent weather reporting. Telephone companies, along with Internet services, can supply up-to-the-minute temperatures and forecasting services. There are exclusive weather-only broadcast frequencies on the radio and television. Buildings post large digital temperature readings for all to see. In Baker, California, a hamburger restaurant claims to have the world's largest thermometer, smack in the middle of the desert for all to come and see. A scorching 118 degree F day in July 1998 brought hundreds of tourists to the restaurant to take photographs of the record-setting temperature display. According to the manager the hotter it got, the more customers arrived; and business rose along with the temperature increases. And as the patrons consumed more and more hamburgers, they added to the list of ways the heat can make you sick—overeating. Overeating when the temperatures are overheating can lead to dire consequences since the normal function of the heart and intestines is disrupted.

TEMPERATURES (HOT)

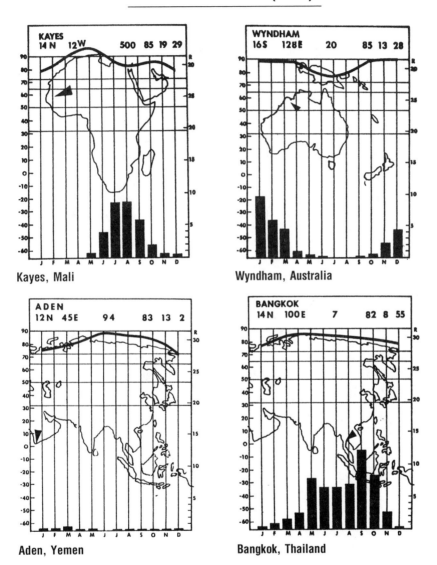

Kayes, Mali

Wyndham, Australia

Aden, Yemen

Bangkok, Thailand

Sweltering heat waves can cause a wide array of adverse impacts on human health and the nature of people's activities. They contribute to the rise in sickness, lethargy, the spread of disease and deaths, especially among the elderly. They place a huge undue strain on a region's infrastructure and essential facilities. People living in perennially hot temperature locations such as those represented here (see key in the Introduction), have, over the years, learned a myriad of ways to "beat the heat."

SOURCES

Clarity, James F. "The Biggest Heat Wave in Years Suns Irish." *New York Times*. July 2, 1995.

Cushman, John H. "Report Says Global Warming Poses Threat to Public Health." *New York Times*. July 8, 1996.

Farah, Mounir A., et al. *Global Insights*. New York: Glencoe, Macmillan/McGraw-Hill, 1994.

"India's Doctors Warn: Stay Out of Record Heat." *Home News Tribune* (East Brunswick, N.J.). May 30, 1998.

Myerson, Allen R. "Heat Wave in Texas Brings 15th Straight Day Over 100 Degrees." *New York Times*. July 21, 1998.

Natural Disasters of North America (supplement). *National Geographic*. July 1998.

National Oceanic and Atmospheric Administration. *Heat Wave*. Washington, D.C.: NOAA, 1994.

Pearce, E. A., and C. G. Smith. *The World Weather Guide*. London: Hutchinson, 1984.

"Rest, Slowdown Advised as Heat Wave Continues." *Home News Tribune* (East Brunswick, N.J.). July 22, 1998.

Saunders, Carol Silverman. "Too Much Sun and Heat Can Mean Trouble." *Current Health*. May 2, 1995.

Sengupta, Somini. "More 90-Plus Heat Expected; Stay Indoors, Officials Warn." *New York Times*. July 15, 1997.

"Southwest U.S. Feeling the Heat as Deadly Temperatures Continue." *New York Times*. July 18, 1998.

Stevens, William K. "Experts on Climate Change Ponder: How Urgent Is It?" *New York Times*. September 9, 1997.

Stevens, William S. "Across a Parched Land Signs of a Hotter Era; Heat Wave's 10 Day Total: 133 Deaths." *New York Times*. August 1, 1999.

Terry, Don. "Heat Death Toll Rises to 436 in Chicago." (Toll raised to 733 by September 1, 1995.) *New York Times*. July 20, 1995.

———. "U.S. Agents in Chicago Track a Subtle Health Hazard: Heat." *New York Times*. October 3, 1995.

Touer, Mike. "Worldwide Warning." *Atlanta Journal Constitution*. January 6, 1996.

The World Almanac, 1999. Mahwah N.J.: Primedia Reference, 1999.

19

TORNADOES

Tornadoes are sudden windstorms that are extremely violent, consisting of cyclonic whirling movements that have a history of causing hundreds of thousands of deaths throughout the world. Tornadoes occur along an abrupt cold front where air masses of strongly contrasting temperatures collide. They usually take place in late spring or early summer on a calm, hot, humid day. Within that mix there is usually a cumulonimbus cloud and a very low air pressure extending downward in a cone-shaped movement of air in a counter-clockwise direction.

In the United States dozens of deaths from tornadoes occur each year, on average forty-four. The country experiences 163 tornado days and 638 tornadoes annually. As many as 1,109 tornadoes were reported in the United States in 1973, a record year. During a seventy-year-period, 1925–1995, as many as 4,944 people died from tornado disasters in the United States. From 1967–1996, death from tornadoes averaged 70 each year.

The word tornado is of Spanish origin and is related to the English word "turn." Tornadoes, popularly referred to as twisters, or cyclones, can readily be spotted as a funnel cloud of small diameter, having wind speeds that can reach incredible velocities of 318 miles per hour (most are 100–150 mph). They normally average about 300–400 yards in width (some could extend a mile or more), travel haphazardly along paths from several miles to fifty miles, most often in a southwest to northeast direction and are associated with very low barometric pressure readings and heavy rains.

HIGH-RISK AREAS FOR TORNADOES

Killer tornadoes usually strike across the plains of Texas, Oklahoma, and Kansas late in winter and early spring. This region is known as "Tornado Alley," but other areas of the United States have also been ravaged by lethal tornadoes. When it comes to where a tornado may hit,

Aerial view of some of the damage caused by the tornado that struck at Lubbock, Texas, on May 11, 1970. (Courtesy of American Red Cross.)

expect the unexpected, it is said. Lethal tornadoes have ravaged the South, the East, and even the northern tier of the United States.

East Pakistan, now Bangladesh, and India have been subjected to devastating cyclonic storms that triggered colossal tidal waves, inflicting enormous numbers of casualties upon the subcontinent. As many as 266,000 to 300,000 tornado-related deaths occurred in the Bay of Bengal region in 1970. Another storm there took 131,000 lives and caused almost $3 billion in damages. In May 1996, still another ferocious tornado tore through eighty villages in a half hour, causing more than 600 Bangladesh villagers to succumb.

Twisters tend to strike in pairs within a short period of time in nearby locations. That is what happened along India's east coast the end of March 1998, resulting in more than 200 deaths, with more than 500 trapped under the debris of collapsed houses. More than 10,000 were left homeless. Three months later another deadly cyclone swept through portions of the coastal port cities of western India, killing at least 415, yet 10,000 and up to 14,000 Indian salt workers may have disappeared without a trace during that catastrophe. Some people are able to flee, never to return to their native villages. Countless others are swept out to sea or are unaccounted for. Unidentified victims are quickly burned by ceremonial custom. Many remain underground.

HISTORIC TORNADOES

Over the years the United States has had few respites from destructive tornadoes. Going back to May 7, 1840, a wicked tornado killed 317 people in Natchez, Mississippi. On March 23, 1913, a severe tornado ripped through Omaha, Nebraska, taking ninety-four lives and doing $35 million in damage. The worst U.S. tornado on record occurred on March 18, 1925. It cut a swath of annihilation along a 219-mile path through Missouri, Illinois, and Indiana. It was on the ground for three hours, a most unusual duration, leaving 689 people dead.

As many as 800 may have perished in a series of tornado outbreaks in seven southern states on February 19, 1884. The total of deaths was unconfirmed. On April 21, 1967, as many as eighteen tornadoes swept through five Midwestern states, killing sixty and injuring over 9,000. In March 1984 as many as thirty-six separate cyclonic storms in the Carolinas killed seventy people. A series of tornadoes struck a fifty-mile-wide section of central Florida on February 23, 1998, killing forty-five Floridians and injuring 260, many of whom lived in mobile home parks, which are especially vulnerable to destruction by powerful tornado winds. On March 20, 1998, a tornado tore an eleven-mile gulch through rural northeast Georgia, bringing havoc, death (thirteen died), and injury to scores more. Like most other tornadoes it struck without any warning. Winds of 260 miles an hour brought a savage tornado to Jarrell, Texas, killing twenty-seven people on May 28, 1998.

On May 3, 1999, an F-5 tornado, the strongest classification of tornadoes, roared across Oklahoma and southern Kansas, causing forty-seven deaths, injuring more than 500, and destroying more than 7,500 buildings. It turned out to be America's most costly tornado—an amount estimated to be over $1 billion. On average only one a year at this magnitude strikes the United States. Wind speeds that range between 261–318 miles per hour will damage strong frame houses, lift dwellings off foundations, and carry automobiles through the air for more than sixty yards. Trees are totally debarked and steel-reinforced structures can suffer damages similar to a bomb explosion.

EFFECTS OF TORNADOES ON DAILY LIVING

People who have witnessed a tornado or have been injured in its wake often suffer life-long emotional scars that stem from their experience. In press or television interviews, they shared their very traumatic, yet vivid descriptive impressions that reveal how terrifying a tornado can be. They tell how it sounds like a runaway freight train coming straight at them.

They compare its arrival out of nowhere to a horrible shrieking noise, or like being in a meat grinder, or like what they imagined to be a bomb blast. Some were sure they were about to die. Many prayed.

Most reported the sounds of broken glass and rattling of windows and doors. Hearing and seeing flying debris, even slabs of concrete, striking homes and objects and viewing the sudden disintegration of buildings were deeply imbedded in the victim's psyche. Many were treated for shock at the time and still must receive medical attention, years after the experience.

Tornadoes appear to be able to produce some amazing miracles that have no rhyme or reason for their actions. They have been known to hop-scotch over adjacent neighborhood homes, only to strike at one or another causing one to crumble in shambles and another remaining intact. Once, in 1918, a check flew 225 miles away from Stockton, Kansas, to Winnetoon, Nebraska, a record distance for a tornado's flying object. Tornadoes have blown cast-iron kitchen stoves away, lifted a steam locomotive from its tracks and deposited it down again square on an adjoining pair of tracks. In 1931 a five-coach railroad train was picked up while crossing Minnesota and hurled into a ditch. Many freakish incidents have been reported of homes being totally destroyed only to find people being saved after taking refuge in bathtubs, closets, or under mattresses or being blown almost a mile away and landing unscathed. There have been reports of clothing being ripped off people, of straws being driven into tree trunks, of 700-pound refrigerators found intact three miles away. During a tornado in Corey, Louisiana, January 18, 1973, a baby was transported by high-speed winds more than three hundred yards, receiving only minor injuries. Similarly, in Concord, Alabama, on April 10, 1998, tornado winds snatched a seven-month-old baby out of her mother's arms. The baby, though seriously hurt, lived. Apparently the all-powerful grip of a mother was no match for the furious wrath of a tornado.

PREDICTING AND PREPARING FOR TORNADOES

The number of tornadoes sighted each year in the United States has tripled in the past half century. In order to minimize tornado destruction, the weather service advises that specific precautions be taken when a tornado is detected by Doppler radar or when a watch or warning is issued. Generally a tornado watch is a kind of alert that conditions are right for tornado formation. The tornado warning gives notice that the tornado is imminent and to take shelter. At home it is a good idea to move to the smallest interior room on the lowest floor. The basement level is a good place to "hide." Closets and bathrooms work well. Also, consider taking refuge under heavy furniture, such as a strong table, in the center of the house. Stay away from windows.

If one is in a car when a tornado is sighted, it is wise to abandon the vehicle and seek shelter in a sturdy building. Do not try to outrun the storm. Cars can easily be overturned or blown away. Diving into a ditch or ravine could also offer some protection. Sometimes a tornado occurs over water. It is called a waterspout, kind of a tubelike spray. If confronted by this type of menacing funnel, quickly head at a 90 degree angle away from the direction of the oncoming threat.

When the infamous May 3, 1999, fierce line of storms swept across the Oklahoma City metropolitan area, many residents of the region credited television weather coverage, prior tornado preparation education, and the availability of underground shelters with saving their lives. When a tornado smashes into a built-up city area the casualties can be enormous. In the southern Great Plains people are gaining an acute awareness of the periodic threats and potential disasters that may result from tornadoes.

Meteorologists track weather systems with Doppler radar equipment and can tell people in the path of a tornado exactly where it is and where it is going, block by block, street by street as the twister approaches their homes. Many then are able to retreat to "safe" quarters or get underneath mattresses in home hallways, closets, and bathrooms. After decades of school instruction, public seminars, and media programs, people have learned to understand storms and know what to do when confronted with one. Yet, no plan is absolutely safe and tornado fears for many tend to be indelible.

THE VULNERABILITY OF MOBILE HOMES TO TORNADOES

A special concern for the potential ruin that tornadoes might bring lies with the residents of mobile home parks. Since mobile homes and recreational vehicles are made of lightweight aluminum, often with extended sixty-foot sidewalls and lack of a solid foundation, they can readily be hurled hundreds of feet into the air resulting in huge damage and much greater loss of life than with more secure conventional dwellings. Since there are approximately 700,000 such mobile dwellings in Florida alone, it is no small wonder that so many were destroyed in the spring of 1998, killing twenty-eight people. Seeing mobile homes flying through the air is a different kind of "mobility," obviously unanticipated and unwanted by mobile-home owners.

KILLER STORMS

Deadliest tornado outbreaks in the United States:

- May 31, 1985—90 killed in Ohio, Pennsylvania, and Ontario.
- April 3–4, 1974—more than 300 killed in "super outbreak" of 148 tornadoes in 11 midwestern states.
- February 21, 1971—110 killed in Mississippi Delta region.
- April 11, 1965—271 killed in Indiana, Illinois, Ohio, Michigan, and Wisconsin.
- May 25, 1955—115 killed in Kansas, Missouri, Oklahoma, and Texas.
- June 9, 1953—90 killed in Worcester, Massachusetts, area.
- June 8, 1953—142 killed in Michigan and Ohio.
- May 11, 1953—114 killed in Waco, Texas, area.
- March 21, 1952—208 killed in Arkansas, Missouri, and Texas.
- April 9, 1947—169 killed in Texas, Oklahoma, and Kansas.
- June 23, 1944—150 killed in Ohio, Pennsylvania, West Virginia, and Maryland.
- April 6, 1936—203 killed in Gainesville, Georgia.
- April 5, 1936—216 killed in Tupelo, Mississippi.
- March 21, 1932—268 killed in Alabama.
- March 18, 1925—689 killed in Missouri, Illinois, and Indiana.
- May 18, 1902—114 killed in Goliad, Texas.
- May 27, 1896—300 killed in Missouri and Illinois.
- February 19, 1884—more than 800 killed in Mississippi, Alabama, North and South Carolina, Tennessee, Kentucky, and Indiana.
- May 6, 1840—317 killed in Natchez, Mississippi.

TORNADOES

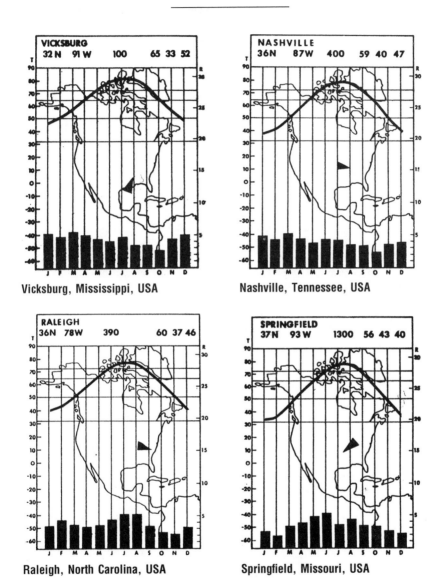

Vicksburg, Mississippi, USA

Nashville, Tennessee, USA

Raleigh, North Carolina, USA

Springfield, Missouri, USA

Tornadoes, the sudden extremely violent twisting windstorms, are generally found in humid subtropical or mild winter climates, where the seasons are usually moist and the summers are long and hot, such as those represented by these climographs (see key in the Introduction). Their incredibly high wind speeds have the capacity to level everything in their paths. Thousands of people have died from tornadoes along their unpredictable, haphazard routes of destruction. Tornadoes have been unwelcomed visitors to these climograph sites in the past.

SOURCES

Bragg, Rick. "Fierce Wind and Rain Tear Up Several States, Killing at Least 35." *New York Times*. March 3, 1997.

———. "Stunned Residents of Tornado-Torn Towns Face Unprecedented Rebuilding." *New York Times*. April 11, 1998.

Fisher, Robert Moore. *How to Know and Predict the Weather*. New York: New American Library, 1953.

Holford, Ingrid. *Weather Facts and Feats* (2nd ed.). Middlesex, England: Guinness Superlatives Limited, 1982.

Jennings, Gary. *The Killer Storms: Hurricanes, Typhoons and Tornadoes*. Philadelphia: J. B. Lippincott, 1970.

Kahl, Jonathan. *Storm Warning: Tornadoes and Hurricanes*. Minneapolis: Lerner Publications, 1983.

McDonald, Kim A. "Tornado-Chasing Scientists Use New Techniques to Probe the Origins of the Deadly Storms." *Chronicle of Higher Education*. July 12, 1996.

Marshall, Tim. "A Passion for Prediction: There's More to Chasing than Intercepting a Tornado." *Weatherwise*. April/May, 1993.

Natural Disasters of North America (supplement). *National Geographic*. July 1998.

Navarra, John Gabriel. *Atmosphere, Weather and Climate: An Introduction to Meteorology*. Philadelphia: W. B. Saunders Company, 1979.

Schmieder, Allen A., ed. *A Dictionary of Basic Geography*. Boston: Allyn and Bacon, 1970.

Schneider, Stephen H., ed. *Encyclopedia of Climate and Weather*. New York: Oxford University Press, 1994.

Statistical Abstract of the United States, 1998 (118th ed.). Washington D.C.: U.S. Bureau of the Census, 1998.

"Toll in Indian Cyclone Rises to at Least 1,000." *New York Times*. October 9, 1996.

"Tornadoes—Nature's Most Violent Storms." A Preparedness Guide. National Oceanic and Atmospheric Administration, National Weather Service, September 1992.

U.S. Department of Commerce. *Tornado Safety Rules in Schools*. Washington, D.C.: NOAA, 1981.

Verhovek, Sam Howe. "Little Is Left in Wake of Savage Tornado." *New York Times*. May 29, 1997.

20

WINDS

Wind is the movement of air caused by differences in air pressure within the earth's atmosphere. It is largely due to the differences in the barometric pressures in a general area from place to place. The greater the difference, or gradient, the greater the wind speed. Conversely, the wider the spread of pressure readings, the less the wind velocity. Wind, the primary transporter of air masses and precipitation, is an important control in the climates of the earth and in determining local weather conditions from day to day.

EFFECTS OF WIND ON DAILY LIVING

Wind speeds and directions have always played significant roles in determining various aspects of history and human geography. They have been prime factors in the sea-age era of exploration and discovery of new lands. They have contributed to the outcomes of military battles. Strong wind conditions have caused deaths and injuries on land, sea, and in the air. High wind velocities have delayed landings of spacecraft. Wind is a key consideration in making navigational plans: Pilots know that it takes less fuel and time to fly from west to east with a tailwind than from east to west against a headwind. These tend to be the prevailing wind directions in the middle latitudes.

Winds have been welcomed when they serve to drive away smog and atmospheric pollutants or bring gentle seabreezes on shore, but unwelcomed when they impact upon the quality of play at a sporting event, or when they dismantle buildings or spread fires. For many centuries, prior to the advent of rural electrification, wind has been used to power windmills, enabling the grinding of grain or the pumping of water. They were first used in Persia in A.D. 644.

Winds have always been a driving force for sailing vessels. Experienced sailors know how to catch a favorable wind so as to maneuver

Windstorms can be extremely damaging. The force of the wind was powerful enough to uproot and blow over a large oak tree, dissecting the roof of a home in Bethesda, Maryland, in 1980. (Courtesy of National Oceanic and Atmospheric Administration.)

their ship (or boat) by tacking (sailing against the wind by a series of tacks, or zigzag movements). In bygone days upon the high seas, ship captains were very familiar with the directions of the prevailing winds. They did their utmost to avoid the doldrums, which are the equatorial ocean regions noted for dead calms and light fluctuating winds. That zone was difficult to cross. For the ancient seafarer, as luck would have it, sailing into the horse latitudes (two latitudinal belts 30 degrees north to 35 degrees south latitude) could bring disaster as well. This area becalmed by light winds received its name when Spanish galleons bound for the New World laden with cargo and horses were often forced to dispose of both in order to lighten their loads so as to be able to sail. If the ships were halted or floundered for weeks or longer, extreme thirst, starvation, and even death of some of the crew could result. Under such maddening circumstances, it was not uncommon to discover that sailors frequently chose suicide by jumping into the sea. Even today people who become sluggish, inactive, and tend to stagnate are said to be "in the doldrums."

DANGERS OF THE WIND

A different kind of wind situation, a gale, with strong winds ranging in speed from 32 to 63 miles per hour, can be equally disastrous to sailors. Many a gale have driven ships off their courses, sometimes onto the shore, or more likely might cause a shipwreck out at sea.

Another kind of problem relating to wind and transportation is turbulence, the violent, irregular motion of air. Here the natural wind becomes a succession of gusts and lulls and with random ups and downs and sideward buffeting. The turbulence is caused by conflicting wind currents and disparities in speed. Aircraft aloft are potentially subject to turbulence. From 1988 to 1998, turbulence in flight has caused about 500 major and minor injuries according to the Federal Aviation Agency. Deaths on jumbo jetliners, some at high elevations, have resulted from massive air turbulence. Often when people get hurt as an aircraft gets hit by heavy turbulence, the pilot is compelled to land for the purpose of getting medical attention for passengers injured by the sharp drop of the airplane. When people fail to heed instructions to fasten their seatbelts, they can readily be slammed into the ceiling of the plane and suffer extensive injuries during a turbulent encounter.

Pilots, when informed about turbulence in advance, may be able to alter their flight plans as to direction and elevation, but still turbulence can happen suddenly and unexpectedly. Pilots, in takeoffs or landings, can also be confronted with a very precarious wind shear, which can quickly cause their aircraft to swerve off course so abruptly that the plane dives and crashes. A wind shear is a condition in which the speed or direction of the wind changes abruptly, thereby endangering aircraft.

Winds can contain chemicals that can travel for thousands of miles, bringing their harmful effects to people living far away from the original source of the pollutants. In March 1999 it was detected by university researchers that the air people breathe in Seattle, Washington, had ten instances of airborne chemicals that spewed from a factory in China.

The most harmful wind of all is a wind that affects people directly downwind from an atomic test fallout, such as those conducted in Nevada in the 1950s. This kind of exposure to radiation, it is believed may cause cancer. Particularly worrisome and of international concern are the children exposed to iodine fallout from the Chernobyl nuclear plant accident in the former Soviet Union in 1986. The very sad result is that they have high rates of thyroid cancer. Winds carried the fallout far beyond the Ukraine where the explosion took place.

Wind flow, depending on source, direction, and velocity, can be a very negative factor in its effect on people. Windstorms have been blamed for highway accidents, downed powerlines and electrical outages, toppled trees and slowdowns in businesses, especially where food or goods are offered at outdoor establishments. High winds have caused boats to sink and impeded rescue attempts. They contribute to the very dangerous riptides at ocean beaches, when swimmers are pulled out to sea by the undertow, causing panic and possibly drowning. Winds also cause ocean water temperatures to be unseasonably warm or cold at certain times along the shore.

Winds that are very hot and dry that blow in from the desert in the fall, like the Santa Ana winds of Southern California, can bring extensive wildfires, health alerts, and medical emergencies. Santa Ana winds can blow through narrow valleys at 35 miles per hour with speeds of gusts to 100 miles per hour. The extremely hot dry south sirocco winds that blow across North Africa become unbearably sticky once they reach the Mediterranean Sea. On April 15, 1997, in Mecca, Saudi Arabia, high winds fanned the flames of a fire that swept through a crowded religious encampment of tents in 104 degree F heat. More than 300 died and 1,290 people were injured.

USING THE WIND

Weather vanes are flat pieces of metal set up high to swing with the wind and show the direction from which the wind is blowing. Weather vanes have always been important weather instruments used by farmers, sailors, pilots, and others concerned with weather forecasting and navigation. Across Europe where wind power is booming with the advent of improved technology, wind turbines increasingly dot the landscape. Wind turbines are engines driven by the pressure of air against the curved vanes or plates of a wheel fastened to a driving shaft.

A number of governments are now encouraging the growth of wind energy because it is cheap to produce and reduces pollution from oil and coal-fired electric energy plants. Besides, the expansion of nuclear plants is hardly being endorsed by environmentalists and others worried about radiation problems. In Denmark there are currently 4,700 modern windmills. Wind power there generates 6 percent of the country's electricity. That is the highest per capita output of wind energy in the world. Wind power is inexpensive to produce, but a near steady wind of 12 miles per hour is required for efficiency. Wind turbines have become quite apparent in the desert area of California. Near Palm Springs 4,000 colossal windmills produce electric power for the region.

Imaginative and inventive people have always found ways to use the force of wind to their advantage. They have developed personal wind turbines for their homes, wind socks to show wind direction at small air strips, weather balloons, and "windcatchers." These devices used on some homes in Pakistan are placed on the roof where they trap wind and funnel the cooling airflow into the house in a kind of natural air conditioning system. And, for the sports minded, kite flying, popular in the Orient, hot air ballooning, and sail boats with their ever changing improvements in design and craftsmanship compel the sports person to be "wind conscious."

WINDS HAVE "PERSONALITIES"

Winds can be calm, slight, gentle, or strong. They might register gale or hurricane force. Each can be innocuous or extremely damaging. Each type of wind, be it the sound or force, can affect people's mood and demeanor. An old adage best describes how winds may bring despair or delight: "There is never an ill wind that does not blow well for someone." A wind that blows away the roof of one man's home may prove profitable for the roofer contracted to repair it. Some winds that blow away topsoil may bring financial ruin to a farmer, while winds that carry very fertile loam soils, called loess, that are deposited on another farmer's acres allow him to increase his crop yield.

WINDS

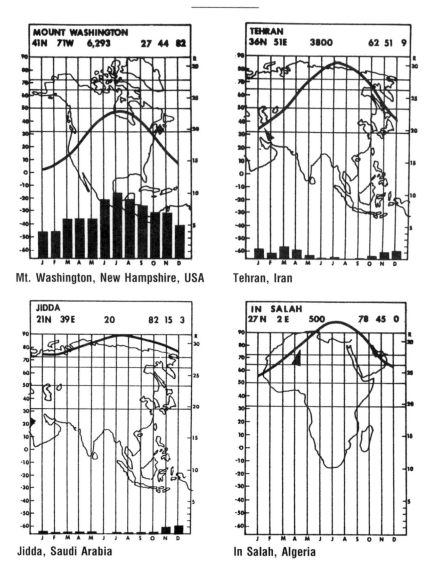

MOUNT WASHINGTON
41N 71W 6,293   27 44 82

Mt. Washington, New Hampshire, USA

TEHRAN
36N 51E 3800   62 51 9

Tehran, Iran

JIDDA
21N 39E 20   82 15 3

Jidda, Saudi Arabia

IN SALAH
27N 2E 500   78 45 0

In Salah, Algeria

Mt. Washington, New Hampshire, is thought to be the windiest place in the United States. The average wind velocity there is 35.3 miles per hour. The greatest wind speed recorded was a high of 231 miles per hour. Elsewhere, such as the other areas represented in the climographs (see key in the Introduction), winds can bring both beneficial and harmful results to specific areas depending upon their speed, source, and direction. Windstorm damage is a bane to homeowners. Winds have played a significant role in shaping certain historical events and are instrumental as a source of power. They also must be taken into account in the navigation of many means of transportation. Wind-borne pollutants, traveling aloft to places far from their originating sources, can have a very harmful effect on the health and comfort of millions of people in their path.

SOURCES

Aquilera, Mario C. "Inventor Develops Wind Turbine for Homes." *Home News Tribune* (East Brunswick, N.J.). December 8, 1996.

"Atomic Tests Fallout May Have Reached East Coast." *Home News Tribune* (East Brunswick, N.J.). July 26, 1997.

Cunningham, William P., et al. eds. "Wind Energy." In *Environmental Encyclopedia*. Detroit: Gale Research, 1994.

Edinger, James G. *Watching for the Wind: The Seen and Unseen Influences on Local Weather*. New York: Doubleday, 1967.

Eldridge, F. R. *Wind Machines* (2nd ed.). New York: Van Nostrand, 1980.

"Fire in Mecca Kills Pilgrims in Tent Camps; 300 May Have Died." *New York Times*. April 14, 1997.

Gipe, Paul. *Wind Power for Home and Business*. New York: Chelsea Green, 1993.

Goldberg, Carey. "California Fires Scorch 35,000 Acres." *New York Times*. October 23, 1996.

Halacy, D. S., Jr. *Earth, Water, Wind and Sun: Our Energy Alternatives*. New York: Harper and Row, 1977.

Horstmeyer, Steve. "Tilting at Wind Chills: Is Winter's Popular Index Blown Out of Proportion?" *Weatherwise*. October/November 1995.

National Disasters of North America (supplement). *National Geographic*. July 1998.

"Pollution from China Detected in Northwest." *Sun-Sentinel* (Ft. Lauderdale, Fla.). March 3, 1999.

Posey, Carl A. *The Living Earth Book of Wind and Weather*. Pleasantville, N.Y.: Reader's Digest Association, 1994.

Schneider, Stephen H., ed. *Encyclopedia of Climate and Weather*. New York: Oxford University Press, 1996.

"The Short Dangerous Life of a Waterspout." Geographica. *National Geographic*. December 1995.

Simons, Marlise. "Across Europe, a Tilt to Using the Wind's Power." *New York Times*. December 7, 1997.

"The Strongest Wind on Record." *New York Times*. December 18, 1997.

"Winds May Compel Balloonist to Fly over Libya." *Sun-Sentinel* (Ft. Lauderdale, Fla.). January 3, 1998.

PART II

HOW WEATHER AFFECTS US

Wind turbines, similar to those found near Palm Springs, California, are being installed all across the United States. They provide a good example of how businesses can benefit from the weather. Wind machines are already supplying economically competitive electricity throughout the world. By the year 2030, wind turbines could generate 10 percent of the U.S. electric power. (Courtesy of Department of Energy/NREL-PIX#00045.)

21

BUSINESS

Weather predicting and weather conditions play a significant role in the world of business. Weather affects 20 percent of the $9 trillion U.S. economy. A commodities trading firm may contract with a weather forecasting service to predict future weather conditions in certain growing regions for the potential harvest of wheat, corn, soybean, coffee, and other agricultural commodities. Based on the predicted supply figures, the company's brokers can advise their clients to buy or sell their animal and grain futures. Frost damage to crops and fruits can send prices tumbling, and a fine growing season of ideal weather can yield an oversupply of farm products, thus lowering prices as well. Rapid weather changes can affect prices of key crops worldwide. In Australia wheat prices dropped 8 percent in one month during 1997 after a heavy rainfall erased the threat of a predicted drought.

An unanticipated winter heat wave can cause a sudden upsurge in tourism, send golfers to driving ranges, customers to ice cream stores, and increase sales by food vendors on city streets. Conversely, ski shops, heating-oil companies, and shops specializing in outerwear, sweaters, boots, gloves, and scarves find business to be slow. Theaters and restaurants see a rise in patrons willing to leave their homes when the weather is less inclement. People get out more to shop in department stores. During snowstorms sales are down as is production in factories due to absenteeism by employees.

The winter season bolsters the sale of snow shovels, snow plows, ice melts, car batteries, snow tires, tire chains, and the like, while summertime brings an increase in sales of lawn care products, swimwear, auto tune-ups, sun lotions, and sunglasses.

California's disproportionate amounts of daylight and bright sunshine hours afforded the impetus of an early start to the movie-making and airplane industries. The mild rather predictable weather proved advantageous for "shooting" at outdoor locations and for long hours of flight training, airplane testing, and storage of newly built aircraft.

An examination of seasonal advertisements in newspapers and magazines will provide readers with a good indication of how closely attuned the business sector of the economy is to climatic conditions and expectations. Retail merchants, in particular, reserve large portions of their advertising for timely promotional announcements about their merchandise prior to and during anticipated weather changes. A store may need to sponsor special end-of-season sales at times when weather uncertainties may lead to disappointing revenues from purchases and cause the manager to try to clear out an oversupplied inventory by offering huge discounts to customers.

Millions of dollars are spent to research and survey ideal sites in which to locate and build tourist attractions, hotels, and theme parks. The places selected, many of which are in Florida and Southern California, are most often those that ensure warm, favorable weather conditions throughout much of the year. The weather factor, is, of course, a prime consideration in the popularity of vacation meccas throughout the world. Rarely, if ever, will investors construct recreational centers where the weather tends to be undesirable for out-of-doors tourism.

The business of farming, its type, success, and nature of the farm operator's methods of working the land, is largely determined by weather elements. In the Great Plains, where wheat is a dominant crop, it takes huge amounts of acreage for a farmer to make a living. The climate is temperate continental with twenty-one inches of rainfall per year, mainly in the summer. The farmer must work the land in a lifestyle engaged in extensive agriculture practices. The farmer of the corn belt, on the other hand, can make a living on much less land. His farming is more intensive. In that region about thirty-eight inches of rainfall is recorded annually, most of it falling in the hot summer months as well. The success of his crop yield is almost entirely dependent upon the timing and amounts of seasonal rainfall, making farming there a risky business.

Owners and managers of construction companies are especially attuned to possible weather outcomes that may slow the completion or halt a job at a site for a number of days due to a storm or inclement weather. Their estimators must anticipate such a possibility when making their job-cost calculations. Losses in time and wages may occur due to extreme cold when it may be unwise to do masonry work or pour certain kinds of concrete. Painters, carpenters, roofers, and iron workers are often unable to perform their tasks, or may be at risk for safety reasons when the weather elements manifest rain, snow, ice, or wind. It is common to have time lost in getting workers and materials to the construction site during times of severe weather.

Companies that manufacture or sell seasonally prone goods are often susceptible to periodic changes in weather norms, which can cut into or increase profit margins. Especially vulnerable to fickle weather results are firms that produce such items as sunglasses, sun lotions, sporting goods, camping equipment, boats and marine engines, bicycles, fungicides, and even beer. Inventories and production schedules become skewed to weather outcomes. Inclement weather of a severe nature can delay shipments. Plant layoffs of workers and retail store closings can result. The stock values of affected companies can readily fluctuate according to weather changes.

Extended periods of rain or a long-lasting heat wave can hamper outdoor fund-raising events and discourage people from purchasing tickets to concerts and sporting engagements in advance. Poor weather conditions can decrease attendance at beaches, parks, pools, and recreational sites, thus hurting sales by merchants and vendors.

Local business operators, service workers, and professionals may find that snow, rain, or windstorms can result in their economic advantage. Icy, slippery, or muddy roads can cause accidents and "fender benders," resulting in people who need repairs to automobiles by body shop owners. Car wash garages do well after streets and highways are treated with chemical salts and sand after snowfalls. Tree service companies are frequently called to remove fallen trees and debris on home lawns. Even orthopedic doctors find their offices very busy during times when walkways become slick. And, of course a sudden rainstorm, hurricane threat, or unanticipated cold spell can drive summer vacationers away from beachfronts en masse. When that happens seasonal profits suffer.

In an interdependent and global economy, the weather of a given place can influence the availability and prices of commodities elsewhere. If an orange grove frost kills a crop in California, or a drought limits the yield of fruits and vegetables in Central or South America, those food items are hard to come by, even at much higher costs. It is no small wonder that when merchants are queried about the weather, a reply may sound something like: "The weather is terrible and so is business!"

BUSINESS

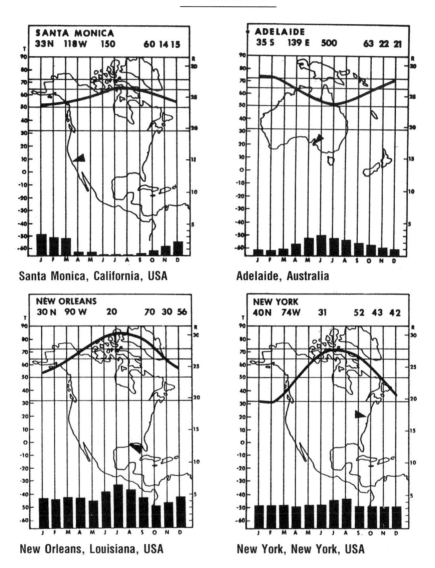

SANTA MONICA
33N 118W 150   60 14 15

Santa Monica, California, USA

ADELAIDE
35 S 139 E 500   63 22 21

Adelaide, Australia

NEW ORLEANS
30 N 90 W 20   70 30 56

New Orleans, Louisiana, USA

NEW YORK
40N 74W 31   52 43 42

New York, New York, USA

All types of businesses are at the mercy or favor of weather conditions. The weather can also be an underlying factor in inducing or deterring trade and commerce. Frequently the climate common to a given place is a major contributor to the various kinds of products or commodities emanating from that region. There is an important interrelationship between weather, business, productivity, and the marketplace. Therefore, businesses may check the climate of major cities, such as these four, when planning their location (see key in the Introduction).

SOURCES

Ayoade, J. O. *Introduction to the Climatology for the Tropics*. New York: John Wiley and Sons, 1983.

Blair, Thomas A. *Climatology: General and Regional*. New York: Prentice-Hall, Inc., 1942.

Burns, Greg. "El Niño Already Casing Headaches; Unusually Warm Water in Stretches of the Pacific Is Worrying Traders." *Chicago Tribune*. September 25, 1997.

"California Climate, on Such a Winter Day." *New York Times*. January 6, 1998.

Farah, Mounir A., et al. *Global Insights*. New York: Glencoe, Macmillan/McGraw-Hill, 1994.

Fisher, James S. *Geography and Development: A World Regional Approach*. New York: Macmillan, 1992.

"Frost Damage Reassessment Sends Wheat Prices Tumbling." *New York Times*. June 12, 1997.

Harper, Robert A., and Joseph P. Stoltman. *World Geography*. New York: Scholastic, 1988.

Miller, E. Willard, and George T. Renner. *Global Geography*. New York: Thomas Y. Crowell, 1958.

"Natural Gas Soars as Utilities Need Fuel Amid Heat Wave." *New York Times*. June 24, 1997.

Resnick, Abraham. *Money*. San Diego: Lucent Books, 1995.

"Soybean Prices Drop Again on Prospects for Record Crop." *New York Times*. July 2, 1979.

Spencer, J. E., and W. L. Thomas. *Introducing Cultural Geography*. New York: John Wiley and Sons, 1973.

Statistical Abstract of the United States 1998 (118th ed.). Washington, D.C.: U.S. Bureau of the Census, 1998.

Stieghorst, Tom. "Quiet Season Calms Insurance Cost." *Sun-Sentinel* (Ft. Lauderdale, Fla.). December 3, 1997.

"Warm Weather Cools Off Kids' Clothing Sales." *Home News Tribune* (East Brunswick, N.J.). October 15, 1997.

The World Almanac, 1999. Mahwah N.J.: Primedia Reference, 1999.

With a temperature of −46 degrees F in some regions of Siberia, proper dress can mean survival. The smallest children are wrapped in layer after layer so that little more than their eyes are visible. (Courtesy of Novosti Press Agency.)

22

CLOTHING

Humans have been able to live in every climatic region on earth. Different kinds of body covering or clothing enable people to maintain healthful body regulation by preventing heat loss in cold weather and by precluding dangerous heat buildup in hot weather. Throughout the history of mankind people have fashioned clothing from available materials made from sheep's wool, fish skins, sea mammals, animal furs and hides, plant fibers like cotton or flax, and silk from silkworms. Most often a garment is made from materials indigenous to an area determined by environment and climate. Now synthetic fibers such as nylon and rayon are used as well. But, they too are selected and worn for specific weather expectations.

Weather and climate set both style and material for dress. Porous, lightweight material, worn loose, shorter, and more open, is designed to allow air to reach the skin in order to cool the body by evaporating perspiration. In cold weather it is better to put on layers of clothes since the cold air is kept out, and the warmth of the body is held insulated between the superimposed clothes. Desert living peoples, like Bedouins and other Arabs, dress sensibly for rapidly changing temperatures and sandstorms. Loose flowing robes allow air to circulate and guard the body against the intense heat of the sun during the day and provide warmth for the huge drop in temperature—as much as 90 degrees in a few hours—after nightfall. An ancient problem is keeping one's feet dry. Wet feet are prone to frostbite and disease, common maladies in cold climate regions. Appropriately insulated footwear are worn for prolonged winter conditions.

Traditional wrap-around sarongs, worn around the lower part of the body as a principal garment of the East Indies by men and women, seem to be appropriate for the hot days of that region. The sari, a long piece of cloth worn wrapped around the entire body in the form of a skirt and draped over the shoulder and head by Hindu women of India also serves as a multipurpose garment for varied weather conditions. The shawl, an oblong cloth worn by some women of Latin America as a covering for their head or shoulders, is a practical piece of clothing always ready for sudden rainstorms or a rapid drop in temperatures. Indians have traditionally relied on draping themselves in blankets for similar relief from the element when needed. And as a very practical protection against rain, the poncho, a cloaklike blanket with a hole in the middle for the head, worn by some people of South America, is an ideal garb for the more rainy areas of that region.

A wide array of weather gear and accessories now comprise suitable options for those inclined to dress for the weather. For the cold there are parkas, hooded down jackets some with fleece or pile lining, boots, ski hats, leggings, gloves and scarves, overcoats and jackets, earmuffs, and even thermal undergarments. For warm days people can don shorts, short-sleeve shirts and blouses, and varied attire made of "breathable" cotton fabric. A wide-brim hat such as a sombrero, popular in Mexico, or a peaked baseball-style cap are always serviceable under sunny skies. The sun's rays can be shielded by sunglasses, or for some a parasol is an option. And for rainy conditions a raincoat, low-cut overshoes, or high-top boots, sometimes called galoshes, might do. They are best for snow or slush. It is always wise to carry an umbrella along—just in case.

Many people are conditioned to subconsciously observe the outdoor weather conditions, or to listen to the daily weather forecasts upon rising in the morning. That will normally dictate the type of clothes to wear for the rest of the day. That was what the people of Rapid City, South Dakota, may have done on January 11, 1911. Since there were no radio weather reports then, their visual observations hardly helped them select the proper clothes for what turned out to be a record-setting change in the weather that day. Apparently Rapid City lived up to its name in an indirect way, for about mid-day within a fifteen-minute span of time, the temperature recorded a "rapid" drop of 47 degrees. The sudden arrival of cold air caused the residents of the city to scurry home and change into much warmer clothing. That implies that in climate areas where the weather is quickly changeable, people need to be ready to shed—or add—clothing that best offers comfort for the moment.

CLOTHING

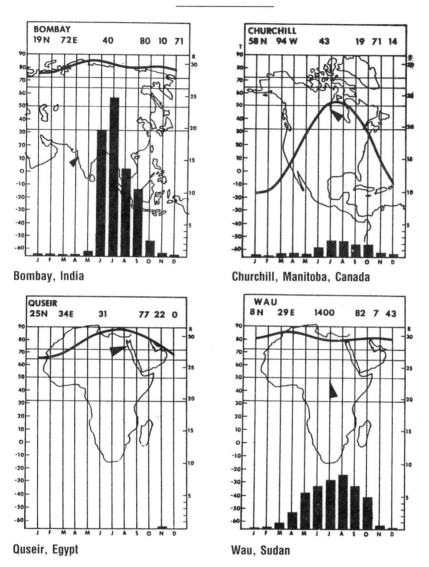

Bombay, India

Churchill, Manitoba, Canada

Quseir, Egypt

Wau, Sudan

Throughout the history of mankind people have fashioned clothing from available materials found in that area, which are often determined by the nature of the climate, vegetation, and animal life associated with the region. The changes in the daily weather dictate the particular gear or type of clothes needed to be worn for that day, causing people to "dress for the weather." People everywhere need to change their garments, shoes, and headwear to meet seasonal variations in temperature. The accompanying climographs best illustrate the wide array of climate requiring clothing adaptations to meet specific local and regional weather features (see key in the Introduction).

SOURCES

Ayoade, J. O. *Introduction to the Climatology for the Tropics*. New York: John Wiley and Sons, 1983.

Carter, George F. *Man and the Land*. New York: Holt, Rinehart and Winston, 1975.

Critchfield, H. J. *General Climatology* (3rd ed.). Englewood Cliffs, N.J.: Prentice-Hall, 1974.

Dicken, Samuel N., and Forrest R. Pitts. *Introduction to Cultural Geography*. Lexington, Mass.: Xerox College Publishing, 1970.

Farah, Mounir A., et al. *Global Insights*. New York: Glencoe, Macmillan/McGraw-Hill, 1994.

Kidwell, Claudia B., and Valerie Steele, eds. *Men and Women Dressing the Part*. Washington, D.C.: Smithsonian Institution, 1989.

Kohler, Carl. *A History of Costume*. New York: Dover Publications, 1963.

Miller, E. Willard, and George T. Renner. *Global Geography*. New York: Thomas Y. Crowell Company.

Oliver, John E., and Rhodes W. Fairbridge, eds. *The Encyclopedia of Climatology*. New York: Van Nostrand Company, 1987.

Racinet, Albert C. A., ed. *The Historical Encyclopedia of Costumes*. New York: Facts on File, 1988.

Resnick, Abraham. *Siberia and the Soviet Far East: Endless Frontiers*. Moscow: Novosti Press Agency Publishing House, 1983.

Spencer, J. E., and W. L. Thomas. *Introducing Cultural Geography*. New York: John Wiley and Sons, 1973.

Schnurnberger, Lynn. *Let There Be Clothes*. New York: Workman Publishing, 1991.

Scott, Stephen. *Why Do They Dress that Way?* Intercourse, Pa.: Good Books, 1986.

Steadman, R. G. "Indices of Windchill of Clothed Persons." *Journal of Applied Meteorology*. August 1971.

23

CRIME

Many criminologists and sociologists for more than 100 years have been doing independent studies in an endeavor to uncover a relationship between crime and the climate. Their research did not find climate to be the principal cause of crime, however, they as well as the U.S. Department of Justice in 1980, linked violent crimes against persons with the heat of the summer months. The findings attributed the eruption of human emotions to the increase of heat during that time of the year. An examination of 40,000 cases of assault and battery occurring between 1891 and 1897 in New York City found that the rise of people's emotions during the hottest times of the summer months led to an increase in fighting.

In the past some eminent researchers maintained that people were more likely to riot or commit assault in May or June, months when insanity and sexual crimes also reach their peaks. Their general conclusion was that other factors besides weather need to be taken into consideration before a definite crime-climate connection can be established. At the very least, they suggest that data indeed show that the warm summer months provide enhance opportunities for the commission of different crimes.

Heat waves, it is speculated, may trigger violent crimes when patience and tempers are short and people become edgy. As a case in point, the murder rate in New York catapulted 75 percent during the heat wave of 1988. The rates of domestic violence also show decisive increases during hot sultry weather. The heat-behavior connection may be indirectly related to the fact that people try to "beat the heat" by consuming larger quantities of cold beer and alcoholic beverages. Hot tempers and hot temperatures cannot be cooled by iced drinks of alcohol.

The Federal Bureau of Investigation in a 1994 report entitled *Crime in the United States* released a number of crime statistics by categories. In summary they found that various kinds of crime were greatest during

Such weather conditions as hurricanes and tornadoes leave merchants vulnerable to looting, as is depicted here in a department store in the wake of Hurricane Luis, which hit St. Maarten in 1995. (*Credit:* AP/Wide World Photos. Reprinted with permission.)

the warmer months of summer and lowest during the colder months of winter. The longer daylight hours of the summer season and the shorter daylight hours of winter were also factors that contributed to the difference in the rates and totals of offenses. In terms of the national crime index, February had the lowest rate for crimes of violence and against property; while the highest total was experienced during August for the same category. The same months were also singled out for the national murder rate, forcible rape (December was slightly less), robbery, property crime, larceny-theft, and motor vehicle theft.

Fairly recent studies highlighted in the *American Journal of Police* confirm a number of linear relationships that connect weather and crime. The frequency and kinds of calls for police services established that (1) crimes in general rise when temperatures become hot, (2) inclement weather, such as cold temperatures or heavy rainfall, reduce the number of potential victims available as people usually stay off the street during bad weather, (3) the summer season affords criminals more opportunities to commit crimes, (4) fewer calls were made during periods of high winds (when disturbing polluted air is dispersed), and (5) crimes decrease when fog is present.

Automobile thefts rise during the summer months when daylight hours are extended and the weather is warmer. During that season more cars are taken from shopping malls, parking lots, and sites away from home neighborhoods. It is noted that there is a higher volume of auto use during times of clear weather. It is also a time when fatalities involving intoxicated or alcohol impaired drivers increase. Holiday periods are particularly troublesome.

In addition to the seasonality of crime victimization as it relates to weather, particularly the temperature factor, one can speculate about the crime patterns and the reasons behind them, which for the most part focus on individual conduct. For example, people tend to spend more time outside in the warmer weather, away from their residences, increasing the potential for home robberies and street crimes. Doors and windows in homes are more apt to be left open and unlocked, creating incentives for intruders to have easy access. More household articles are found to be left on the front lawn or backyard, leading to invitations to theft. People tend to shop more during summer months and may have a tendency to leave merchandise in automobiles as a target for thieves. Also, vacationers are away from their unwatched homes, or staying at resort areas where incidents of crimes are more frequent.

During the winter season, when the weather can be cold and harsh, people tend to be more homebound, perhaps snowbound as well, sitting in front of a cozy fireplace with snow falling outside. What burglar wants to leave his shoeprints in the snow?

Sometimes, in the aftermath of a natural disaster, such as a flood, hurricane, or tornado, when property damage is extensive, human behavior may turn negative. Mobs have been known to break into damaged homes and buildings in order to loot and plunder. This is a highly punishable criminal act that tends to manifest itself worldwide. For some a "finder's keepers" mentality may prevail after a storm. Items strewn about and discovered intact may never be returned to their rightful owners.

A new fascinating specialization is currently emerging and receiving much attention in courts of law. It is called forensic meteorology, which is a study of microclimates and crime. Forensic meteorologists are now being called upon to give expert testimony about weather conditions at the precise time and place where a crime is committed or damages to property occurred. They research and often bring exact data relating to all kinds of weather conditions to a trial that may help determine the outcome of a case. Accidents due to weather situations, which may be at the nub of a dispute, frequently draw the attention of these specialists. Their findings are even being relied upon in murder cases, as Abraham Lincoln once did in refuting a claim by a witness that he clearly saw a murder take place by the light of the moon. Lincoln, by citing the 1857 *Farmer's Almanac*, showed that the moon was in its first quarter and riding low on the horizon at the time of the crime, so that it would have shed almost no light on the crime scene. The defendant was found guilty. Today's lawyers are increasingly relying on weather records as evidence in their presentations to judges and juries.

Over the years a number of countries punished criminals and political offenders by sending them to distant places, usually to perform very hard work in underdeveloped regions. Most often the climate at their points of destination was very harsh. Britain transported more than 160,000 convicts to Australia from 1788 to 1868. For 150 years France sent prisoners to extremely cruel penal colonies in sweltering French Guiana and Devil's Island. During the twentieth century tens of thousands of Soviet convicts and political prisoners were banished to remote outposts in Siberia, called gulags. Located in the frozen swamplands of the Arctic or Taiga, the coniferous forests in the far northern regions of Eurasia, many of the exiles unable to regain their freedom froze there for years, literally to their deaths.

CRIME

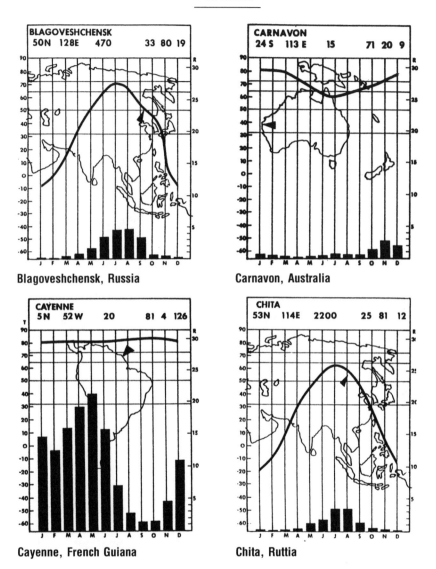

Blagoveshchensk, Russia

Carnavon, Australia

Cayenne, French Guiana

Chita, Ruttia

The kinds of human relationships, activities, and behavioral patterns can indirectly be influenced by the amount of daylight hours, season, and weather conditions found at a specific place and time. They may also have a bearing on the types and incidences of criminal conduct, as reported by law enforcement authorities. Historically, the selection of places to exact criminal punishment for convicts has often been decided by their remoteness and harsh climate conditions, characteristics represented by the areas portrayed in these climographs (see key in the Introduction).

SOURCES

Anderson, C. A. "Temperature and Aggression: Effects on Quarterly, Yearly, and City Rates of Violent and Nonviolent Crime." *Journal of Personality and Social Psychology*, 42. 1987.

Cheatwood, Derral. *"The Effects of Weather on Homicide." Journal of Quantitative Criminology*, 11, no. 1. 1995.

———. "Weather and Violent Crime: A Reply to Perry and Simpson." *Environmental and Behavior*, 22. 1990.

Cohn, Ellen G. "The Effect of Weather and Temporal Variations on Calls for Police Service." *American Journal of Police*, 15, no. 1. 1996.

———. "The Prediction of Police Calls for Service: The Influence of Weather and Temporal Variables on Rape and Domestic Violence." *Journal of Environmental Psychology*, 13, no. 1. March 1993.

DeFronzo, J. "Climate and Crime: Tests of an FBI Assumption." *Environment and Behavior*, 16. 1984.

Falk, G. I. "The Influence of the Seasons on the Crime Rate." *Journal of Criminal Law, Criminology and Police Science*, 43. 1952.

Hwang, Antonio. *Seasonal Indices in the Crime in the United States*. Washington, D.C.: Uniform Crime Reports, Federal Bureau of Investigation. March 14, 1994.

Lab, S. P., and J. D. Hirschel. "Climatological Conditions and Crime: The Forecast Is . . . ?" *Justice Quarterly*, 2. 1988.

LeBeau, J. L., and R. H. Langworthy. "The Linkages between Routine Activities, Weather, and Calls for Police Services." *Journal of Police Science and Administration*, 14. 1986.

Lewis, L. T., and J. J. Alford. "The Influence of Season on Assault." *Professional Geographer*, 27. 1975.

Michael, R. P., and D. Zumpe. "An Annual Rhythm in the Battering of Women." *American Journal of Psychiatry*, 143. 1986.

———. "Sexual Violence in the United States and the Role of Season." *American Journal of Psychiatry*, 140. 1983.

Rotton, J., and J. Frey. "Air Pollution, Weather, and Violent Crimes: Concomitant Time Series Analysis of Archival Data." *Journal of Personality and Social Psychology*, 49, 1985.

U.S. Department of Justice. *Crime and Seasonality*. Washington, D.C.: Bureau of Justice Statistics, 1994.

———. *The Seasonality of Crime Victimization*. Washington, D.C.: Bureau of Justice Statistics, May 1988.

24

CUSTOMS

Most human customs and ritualistic ceremonies have a long history that probably took root centuries ago. They were no doubt started by primitive societies in their attempt to explain the unknown, which were often the mysterious elements of weather, be it rain, snow, wind, thunder, lightning, or the like. The position of the sun and climate changes relating to the arrival and passing of the seasons were particularly mystical, beyond their comprehension. Their beliefs about the supernatural often led to superstitions, some aligned to early religious rites. That is why many customs and celebrations evolved out of weather and climatic foundations.

All over the world people observe special calendar days that recognize seasonal differences. The nature of each observance often depends on the kind of weather experienced at a given place at a specific time of the year. In Russia, for example, where the cold, snow, and ice often linger for long winter months and beyond, the people there are most anxious to bid farewell to their frigid conditions and to greet the coming of springtime. They, therefore, in great anticipation of a warming trend ahead, celebrate a merry seven-day holiday called Shrovetide as early as the first weeks in March. It is a time for fun and frolic, carnivals, last chance snow activities, spirited parties, and particularly the eating of pancakes, symbolic of the rounded sun that would soon fill the air with warmth.

For ancient peoples the shortest day of the year, December 21 in the Northern Hemisphere, the winter solstice, meant that from that day on the sun would grow stronger and the hours of daylight would increase. So people held ceremonies in honor of the sun's eventual return, knowing that the sun would enable them to plant crops for their survival.

For many cultures the impending arrival of spring is celebrated with a theme of a fresh new start. Spring cleaning is commonplace. In many places there are special celebrations for the spring equinox on or around

Many world-wide customs are dependent on specific weather conditions. The Iditarod Trail Sled Dog Race takes place in Alaska, an ideal spot because there is plenty of snow. (*Credit:* AP/Wide World Photos. Reprinted with permission.)

March 21, when the sun is overhead at the Equator. Masked parades are held in the Austrian mountains and in Slovakia, Germany, and Switzerland effigies, or statues of "Death" or "Winter," are burned and dumped in the river.

People the world over still hold on to long-standing local customs inspired by religious traditions and seasonal changes. Colorful Mardi Gras carnivals are held at season's and or start in many Catholic countries of Europe and South America. Ice-carving festivities have become part of wintertime customs in many regions where ice is prevalent. Celebrations in India pay tribute to the arrival of the welcomed rainy monsoon and the end of the drought and dust storms. In the remote Inca villages of Peru, mountain people clad in traditional robes rise to pay homage to the sun during the dawn hours of their winter solstice on or about June 24.

At the first signs of spring a renewal of life is celebrated in Israel with the planting of trees. And for every baby girl born during a holiday called TuB'shvat, a cypress tree is planted. Cedars are planted for the arrival of a boy. The month of August is harvest time on farms in the Northern Hemisphere. Hayrides in horse-drawn wagons are customary. In Britain, a celebration is held to give thanks for the bread made from the year's first wheat. In the United States the Thanksgiving holiday is observed late in November to commemorate the Pilgrim's celebration of the good harvest of 1621.

The beginning and end of winter have a different significance for diverse cultures. In China the winter solstice implies the dying year, thus a feast held at that time may call for empty chairs set aside for the spirits of deceased ancestors. But along the Alpine valleys of Switzerland and Germany, it is customary to have costumed townspeople welcome back their herdsmen and flocks descending from the higher summer pastures into sheltered winter grazing lands by decorating the animals with flowers. And, the universal holiday of Christmas, celebrated during the last weeks of December, when cold temperatures and snow often rule, the warmth of the season is underscored by the festivities that highlight gift-giving, decorating homes and buildings, singing carols, family visitations, well-wishing, and church worship.

One of the most exceptional festival-type activities of all takes place each spring during May in the Kazanluk Valley, a beautiful location within central Bulgaria. Called the "Valley of the Roses," many of the girls and women living there wake up in the middle of the night, dress in regional costume, then gather rose petals before sunrise, while the roses are still moist and oily. The petals eventually supply attar of roses, an oil that is exported and used in making perfumes and an array of other products.

The ability to end a lengthy drought by performing magical rainmaking rites has long been a familiar custom in parts of rural Africa. Some bushmen offer prayers to celestial bodies and are paid for their services in the belief that they can control rain, thunder, and lightning. It is thought that unique powers will bring about an abundance of plant foods, good luck in hunting, recovery from illness, and protection against dangerous animals.

It has been suggested, hopefully in jest, in certain quarters, that some kind of witch doctor now be commissioned to bewitch and charm El Niño, the giant weather maker that has caused major warming of the equatorial Pacific Ocean. El Niño has seriously affected changes in weather patterns throughout the world and has been blamed for all kinds of bizarre climate upheavals. Noteworthy of late are the abnormal rainfall amounts, temperature extremes, crop yield reductions, air pollution, and wildfires that are causing disruptions in ecological balances. Considerable financial troubles, which may have an impact on people's lives and the world marketplace, have resulted from this major weather upheaval. If the mythical witchcraft could by some miracle become a reality, and El Niño's work be brought into line and climate patterns reverted to "normal," everyone might consider a new holiday celebration—"El Niño Day."

The arrival of the Arctic sun is cause to celebrate with a unique ceremony that recalls an ancient tradition for Eskimos living in and around the very remote hamlet of Igloolik, in the far Northwest Territories of northern Canada. In that area, where the temperatures hover about 32 degrees below freezing, the Igloolik people dwell mainly in sod houses and igloos built over perpetual ice. They have survived there, a mere 1,200 miles from the North Pole, for 4,000 years.

At the high-latitude location of Igloolik, the sun can be barely sighted rising during the early winter months. The continuous darkness is only slightly less intense during a three-hour period around mid-day. The isolated and remote little village experiences seven weeks of darkness that locals refer to as the "great darkness." At the rare moment when the sun barely rears itself above the horizon and the darkness fades a bit, the simple ceremony begins. The children who are fortunate enough to catch a glimpse of the sun are rewarded with a special honor. They are chosen to blow out the flames of the soapstone lamps fueled by seal blubber, lit by the area's elders as hundreds gather to view the ceremony. Darkness results for a minute or so until elderly women relight the wick of the ceremonial lamp, which is supposed to represent new life. Though change is occurring rapidly in some aspects of Igloolik life, the people of the region still tend to crave their ancient traditions, like the sun ceremony, seemingly frozen in time.

In the United States February 2 is celebrated as Groundhog Day. A groundhog known also as a woodchuck or a marmot, grows over two feet long and weighs eight to twelve pounds. A superstition that the groundhog can predict weather was brought to America by the British and Germans. They believed that if the groundhog sees his shadow, he is frightened and goes back to hibernate for six weeks; it was believed that this meant more winter weather. An early spring was anticipated if the groundhog did not see his shadow.

This day is also called Candleman's Day, a Christian festival of candle blessing. Punxsutawney and Quarryville in Pennsylvania have Groundhog Banquet and Shadow forecasting festivities on this day. The area's groundhog is called Punxsutawney Phil. He receives much national attention from news and weather reports each year.

CUSTOMS

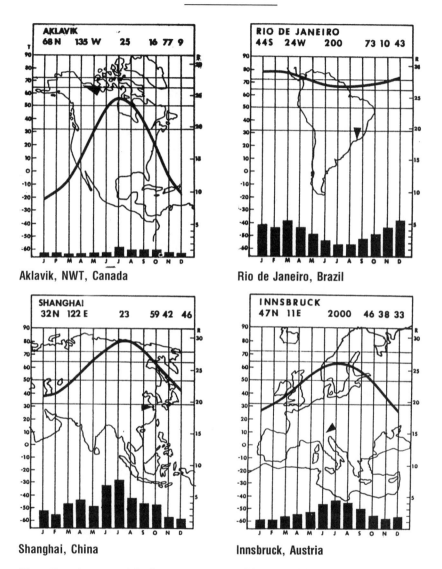

Aklavik, NWT, Canada

Rio de Janeiro, Brazil

Shanghai, China

Innsbruck, Austria

Since time immemorial when man was unable to explain seasonal changes and the weather around him, superstition and speculation ruled many theories about climatic elements. Subsequently, over a period of time, and in diverse locations, local customs and celebrations evolved in an attempt to rationalize the mysteries of weather fluctuations. Today, everlasting weather-related ceremonial-type tributes are still retained in many places worldwide at certain times of the year, including those with the climate extremes represented in these climographs (see key in the Introduction).

SOURCES

Botkin, Ben A. *A Treasury of New England Folklore*. New York: Crown, 1947.

Cohen, Hennig, and Tristram Potler Coffin, eds. *The Folklore of American Holidays*. Detroit: Gale Research Company, 1987.

Cordello, Becky Stevens. *Celebrations*. New York: Butterick Publishing, 1977.

DePalma, Anthony. "How to Love the Arctic Sun? With Ceremony." *New York Times*. January 18, 1999.

Dorson, Richard M. *Buying the Wind: Regional Folklore in the U.S.* Chicago: University of Chicago Press, 1964.

Douglas, George William. *The American Book of Days*. New York: H. W. Wilson Company, 1961.

Dunkling, Leslie. *A Dictionary of Days*. New York: Facts on File, 1988.

Eichler, Lillian. *The Customs of Mankind*. New York: Doubleday, Doran, 1924.

Farah, Mounir A., et al. *Global Insights*. New York: Glencoe, Macmillan/McGraw-Hill, 1994.

Green, Thomas A., ed. *Folklore: An Encyclopedia of Beliefs, Customs, Tales, Music and Art*. Santa Barbara, Calif.: ABC-CLIO, 1997.

Hand, Wayland, Anna Casetta, and Sondra B. Thiederman. *Popular Beliefs and Superstitions, A Compendium of American Folklore* (Vol. 2). Boston: G. K. Hall and Company, 1981.

McDonald, Margaret Read. *The Folklore of World Holidays*. Detroit: Gale Research, 1992.

Meyers, Robert J. *Celebrations*. New York: Doubleday, 1972.

Resnick, Abraham. *Bulgaria*. Chicago: Children's Press, 1995.

Resnick, Abraham, and Margaret Pavol. *Every Day and Every Way*. Bridgewater, N.J.: Far Scribe Educational Publishers, 1988.

Resnick, Abraham, and Margaret Pavol, and Helen Pappas. *Every Day's a Holiday*. Carthage, Ill.: Fearon Teacher Aids, 1991.

Tuleja, Ted. *Curious Customs*. New York: Harmony Books, 1987.

25

HEALTH

It seems that most people respond positively to queries about their well-being when weather conditions are favorable. When the weather factor tends to be depressing, so too goes their mood and sense of good feeling. It may be that this is an age-long truism, perhaps dating back more than 2,000 years when a very wise Greek physician named Hippocrates uncovered a link between daily and seasonal weather and human health and disease. That is one of the reasons why historians regarded him as the "Father of Medicine." Since his early findings, countless investigations and medical studies have confirmed his conclusion that weather and health are interrelated.

Today medical experts take into account the important effects of weather in making diagnoses and in prescribing therapy for their patients. More than ever they have become conditioned to the harmful ways extreme heat or cold, air pollen, the sun's rays, wind, humidity, or smog can be underlying causes of illness. Such chronic and acute ailments, such as respiratory disorders, lung problems, heart failure, heat exhaustion, and heatstroke, may be weather related, particularly during heat waves. During excessive cold spells physicians treat many more cases of respiratory infections, namely pneumonia, bronchitis, and influenza. Surprisingly, medical research has found the "common" cold is not caused solely by people "catching" it during periods of bad weather (a sudden drop in temperature) but more likely by an infection brought on by a virus when an exposed person is in a receptive state. People experiencing fatigue, especially during winter months while working indoors close to others, are more susceptible to getting a cold from a germ-carrying colleague they may have contact with.

According to physiological studies, children tend to be less affected by hot and humid weather than adults. Older people are more lethargic, edgy, and impatient during climatic discomforts. They also seem to be less alert, slower in judgment, and less tolerant during times when they "feel" the weather.

The U.S. Coast Guard is required to carry out many kinds of operations to ensure the health and safety of citizens during inclement weather conditions. Here, their mission is to rescue a family from rapidly rising flood waters, which can cause sickness and epidemics. (Courtesy of U.S. Coast Guard.)

Medical doctors, physiologists, and disease control researchers, using data from patient histories and scientific investigations, have been able to draw meaningful generalizations about the relationship of health and weather in terms of cause and effect. They have found that during extremes of temperatures—both cold or hot—fatalities due to stroke or heart attacks rise 50 percent above normal. Heat waves in the United States killed 4,600 Americans in 1963 and 1,327 in 1980, all within a concentrated area in a short period of time. In the northern United States the mortality rate of coronary heart disease is considerably higher in January and February and lowest around July and August. Bouts with ulcers tend to be less frequent in midsummer. In the warmer regions of the country more people die during the summer than winter. Hospitals report a dramatic and sudden upsurge of admissions to emergency rooms during extreme temperatures, especially among the elderly and poor. During a ten-day heat wave in June and July 1966 in New York City, one-third of all deaths recorded were deemed heat-related.

There are additional weather-health correlations that tend to arouse the interest of medical and lay observers. Asthma, for example, appears to worsen for sufferers during times when the summer air quality is poor, stagnant, or excessively hot. On the other hand, asthmatic difficulties are said to diminish during the winter and spring seasons. A sudden decisive temperature drop can trigger an asthma attack. Autumn usually brings uneasy times for asthmatics and people prone to hay fever. A temperature of 77 degrees F is optimum for a person's metabolism, when chemical and physical processes are taking place within the body.

Some patients tell their physicians that their chronic ailments like arthritis, gout, and muscular back pains "act up" and may even foretell impending weather changes, particularly a forecast of rain. There seems to be a medical connection between rising humidity, coupled with a falling barometric pressure, and how a person feels, healthwise. Cold wind along with high humidity often causes severe chest pains for people with heart disease if they are outdoors in those conditions. Overweight individuals suffer more from heat, while slender people are most uncomfortable in the cold.

Turbulent winds carry plant pollens and dust into the atmosphere, sometimes keeping allergen substances afloat over an area for days on end, giving hay fever sufferers great misery. However, a few days of rain can wash out the irritants and clean the air of pollutants. When automobile and factory pollutants fill the air at the same time, there is a temperature inversion aloft (warmer air sitting over cooler surface air with little air movement) and a serious smog condition can result. This is what took place in a valley near Donora, Pennsylvania, on October 25, 1948, for six days, killing twenty people and causing 2,000 to fall ill from breathing difficulties and respiratory problems. On November 8, 1953, a ten-day temperature inversion caused a bad smog situation resulting in the death of 200 New Yorkers. The worst smog tragedy took place in England in December of 1952, when smog choked off the lives of 4,000 London residents.

Some who study meteorology and biology contend that the weather factor plays a surprising role in determining the physical and mental health of people. There is a contention that the swelling of feet is usually due to increasing moisture and falling pressure before a storm. That manifests itself on airplanes having less than perfect pressurized cabins. Research reveals that an illness called hypoxia can result when humans lack oxygen at high elevations, leading to drowsiness and slowed reflexes. Researchers of hospital records contend that more babies are born during abrupt changes in weather associated with showers, colder gusty winds, rising pressure, and falling humidity. At the U.S. naval base on Guam in the South Pacific on December 17, 1997, during a wind gust of 236 mass per hour (the highest ever recorded), along with a very low barometric pressure, nine women gave birth in a few hours. School psychologists conclude that behavior problems of children in school become more pronounced when the room becomes hot and sticky. There is also a belief that there is a greater incidence of personality mood swings, and even suicides, when the air is warm and moist, the sky is cloudy, pressure falls, winds pick up, and precipitation is taking place.

There is little doubt that an array of infectious diseases that strike humans and animals are strongly related to the geographical environment, in particular, the climate and vegetation of a place, and are transmitted by insects, such as mosquitoes, ticks, fleas, and other blood-sucking parasites. These carriers inflict disease such as malaria, sleeping sickness, and Chagas' disease, which is a fatal disease that affects large numbers of the rural population in Central and South America. There and in Africa as many as 50 percent of the people of certain areas (primarily the hot, wet tropics) are afflicted by deadly insects. Often the mortality rate of the victims is closely related to humidity and temperature conditions. Because of the insect menace, entire villages in Africa at times need to be abandoned. A horrible disease known as river blindness is spread by the blackfly of West Africa mostly during its wet period. The river waters there allow the flies to multiply by the millions. When the blackfly bites a person, worms develop inside the body, sometimes growing over two feet long, and the end result can be blindness.

In the wake of heavy rains, droughts, and other extreme weather a variety of sicknesses and epidemics can occur. That is why scientists are concerned about the impact El Niño and global warming could have on health. In 1997 two Mexican hurricanes led to the formation of many standing pools of water where disease-carrying mosquitoes bred, eventually causing an outbreak of 10,755 malaria cases. Tens of thousands of people in Kenya and Somalia became sick, over 200 died, after the heaviest rains there since 1961 brought on Rift Valley fever, a mosquito-borne disease. Forest fires in Southeast Asia caused by a killing drought subjected hundreds of thousands of people to respiratory ailments. Rains and floods in Uganda, Mexico, and parts of South America caused hundreds of people to die of cholera, which if untreated can kill within two hours by causing severe diarrhea and vomiting. In the southern Rockies, the warm wet winter of 1998 produced abundant food and cover for deer mice, which transmitted a deadly virus to three people resulting in their deaths.

In hot, tropical Africa, within 10 degrees of north latitude and south of the Equator, a region with fifty-five million inhabitants, the people have a tremendous fear of the tsetse fly, a parasite that bites humans and can cause a dreaded sleeping sickness, which if untreated leads to the worst kind of suffering and eventual death. Epidemics have periodically swept through the region. It is estimated that at least 200,000 died of the illness in the outbreaks between 1915 and 1940. In Uganda, an epidemic killed four million people in 1906. Entire villages have been wiped out by the disease. It is tragic that the drug that can cure those who fall ill is much too expensive for local villagers to afford, which means that thousands of unfortunate Africans are doomed to die unless international health organizations and charities can reach them to administer early treatment. In Uganda the life expectancy at birth in 1995 was only thirty-six for a male and thirty-seven for a female. In the United States for the same year males were expected to live seventy-three years and females eighty years.

When the Panama Canal was being built by 35,000 workers in 1913–1914, their greatest task was to control the terrible tropical diseases found in the isthmus there. The region was known to have the hottest, wettest, and most unhealthy weather conditions in existence in all of Latin America. Yellow fever, malaria, and bubonic plague were rampant there, brought on by mosquitoes and other pests. Thousands of canal laborers and others died in years of futile efforts to complete the project until an American campaign to wipe out the diseases became successful through the introduction of many sanitary measures.

On October 8, 1871, a great fire engulfed the city of Chicago burning it to the ground in one of the worst fires in the history of the world. The summer and autumn had been excessively dry. To make matters worse, a high shifting wind fanned the flames very rapidly over hundreds of acres. In a short period of time 18,000 buildings were destroyed and more than 90,000 people—one-third of the population—were left homeless. Property losses were more than $200 million. It was reported that 250 people died, many from smoke inhalation and related respiratory ailments directly and indirectly related to the infamous Chicago fire.

In the aftermath of Hurricane Mitch, which devastated Central America in late October 1998, as many as 50,000 children were exposed to malaria in Tegucigalpa, Honduras, as well as thousands of others similarly afflicted across the country. Large areas of stagnant water left by the floods became breeding grounds for infection-carrying mosquitoes. Thousands of cases of acute respiratory infections, diarrhea, and cholera swept the region, causing hundreds of deaths, many from unsanitary conditions and contagious diseases.

In another region of South America, within the scorching, rainy, and swampy states of Mato Grosso, in southwestern Brazil, epidemics of dengue fever have occurred. This disease is caused by a virus transmitted by the aedes mosquito, which bites during the daytime. It is so painful for its victims that dengue is commonly called the "breakbone fever." It is sometimes fatal in children.

People can be "weather sensitive." A number of doctors, particularly in Europe, contend that at least one out of every three people should be considered "weather sensitive." This category of patients most often blame weather conditions for their uneasy feelings and atypical state of mind. They readily or excessively complain that their symptoms are brought on by the weather at a particular point in time.

The most common complaints due to the weather were tiredness, irritability, dislike of work, headaches, and disturbed sleep. In addition, those who fault the weather frequently attribute problems such as lack of concentration, nervousness, body pains, visual disturbances, increased forgetfulness, an overall lethargic demeanor, dizziness, and a tendency to making more errors than usual to an awareness of weather and how they are affected by it.

Women, as a group, tend to be more weather-sensitive than men. The same holds true for older people, especially those over sixty years of age. Though most individuals react to weather conditions in various ways, it is safe to say that sooner or later everyone will admit to "being under the weather," be it an ache or pain or a slight feeling of the "blahs."

HEALTH

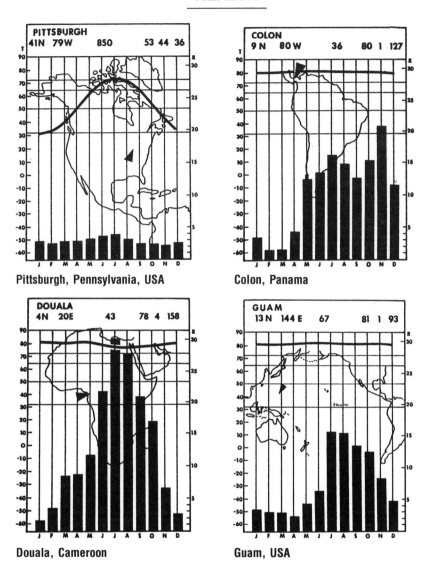

Pittsburgh, Pennsylvania, USA

Colon, Panama

Douala, Cameroon

Guam, USA

Meaningful generalizations about the relationship of health and weather in terms of cause and effect have been made based on patient histories and scientific investigations. When weather conditions become extreme, or atypical, or are subject to sudden change, an array of health problems, even death, can result. Various regions in the world tend to demonstrate commonplace diseases and sicknesses brought on by detrimental climates, such as those depicted in the climographs that have very heavy rainfall or very high or very low temperatures (see key in the Introduction).

SOURCES

Ayoade, J. O. *Introduction to the Climatology for the Tropics*. New York: John Wiley and Sons, 1983.

Carter, George F. *Man and the Land*. New York: Holt, Rinehart and Winston, 1975.

"Cholera Kills 108 in Uganda." *Sun-Sentinel* (Ft. Lauderdale, Fla.). December 25, 1997.

Critchfield, Howard J. *General Climatology*. Englewood Cliffs, N.J.: Prentice-Hall, 1983.

Farah, Mounir A., et al. *Global Insights*. New York: Glencoe, Macmillan/McGraw-Hill, 1994.

Grisham, John. *The Testament*. New York: Doubleday, 1999.

"Illnesses Linked to El Niño Heat." *Home News Tribune* (East Brunswick, N.J.). October 11, 1998.

McKinley, James C., Jr. "Deadly Epidemic Emerges in Sudan." *New York Times*. July 18, 1997.

Miller, Benjamin F., ed. *Family Health Guide and Medical Encyclopedia*. Pleasantville, N.Y.: Reader's Digest Association, 1970.

Oliver, John E., and Rhodes W. Fairbridge, eds. *The Encyclopedia of Climatology*. New York: Van Nostrand Rheinhold, 1987.

Pearce, E. A., and C. G. Smith. *The World Weather Guide*. London: Hutchinson, 1984.

Smith, Stephen. "Mites, Mold Rally Troops during Humid Weather." *Sun-Sentinel* (Ft. Lauderdale, Fla.). August 26, 1997.

Spencer, J. E., and W. L. Thomas. *Introducing Cultural Geography*. New York: John Wiley and Sons, 1973.

Statistical Abstract of the United States 1998 (118th ed.). Washington D.C.: U.S. Bureau of the Census, 1998.

Stevens, William K. "Warmer, Wetter, Sicker: Linking Climate to Health." *New York Times*. August 10, 1998.

"Two Mexican Hurricanes Leave a Legacy of Malaria in Wake." *Home News Tribune* (East Brunswick, N.J.). August 28, 1998.

The World Almanac, 1999. Mahwah N.J.: Primedia Reference, 1999.

26

HISTORY

The importance of the weather factor is usually considered by historians when recognizing causes and results of historical developments. They acknowledge how weather has played a part in determining routes of early explorations, discoveries, settlement locations, migrations of people, past trade relations, and specialized economic patterns worldwide. Weather has had a unique and significant role in shaping the outcomes of battles and military campaigns and in forging national destinies. The weather rationale is also used to explain the basis of how many cultures and institutions, like slavery, started in the United States.

American Indians are basically Mongoloid, though considerable variation is found. It is generally agreed that they migrated into northern America from Asia during prehistoric times of glacial advances when there was a land and ice bridge connecting the two continents at the present site of the Bering Strait, a mere distance of forty-five miles.

Some historians have theorized that the birth and growth of early civilization may have begun along the Nile River oases. It was there that the hot desert climate provided a year-round growing season augmented by the deposit of fresh soils brought to the region from elsewhere by the annual floods. This, in turn, gave rise to early irrigation that bolstered farm yields.

Similar geographic features contributed to an emerging civilization that sprung up around 3500 B.C. on the flood plains of the lower Tigres and Euphrates Rivers in Mesopotamia.

It is theorized that the prevailing winds and currents of the Atlantic Ocean were key factors in determining land claims based on early explorations and discoveries. The trade winds drove sailing vessels westward from the northeast to the West Indies, and they returned to Europe via the steady blowing westerlies. The Norse sailors were delivered to America by the polar gales and northeasterly winds, thus explaining their probable presence in the higher latitudes of North America prior to 1492.

George Washington, in a surprise attack on the British and Hessian troops just north of Trenton, New Jersey, won a military victory that proved to be most memorable and important for his Continental Army. He moved his men across the ice-clogged Delaware River in a driving sleet and snowstorm on Christmas night 1776.

In the 1700s climate played a role in the exploration and occupancy of English, French, and Spanish lands in the new world. Lord Baltimore's colonial effort in Newfoundland was abandoned in favor of the more hospitable weather of Maryland. The Spanish explorations eventually led to settlements in Florida, the Caribbean Islands, and Mexico, where the climate was warm and to their liking. The early French trappers of North America roamed the colder fields and forests of Canada and the low-lands beyond the Appalachian Mountains in quest of fur-bearing animals for their skins.

Rainfall amount, temperature averages, and lengths of the frost-free growing season all entered into the feasibility of growing certain crops in the United States. Early on, in the southern region with months of high temperatures and a plentiful supply of rain throughout the year, tobacco- and cotton-growing became profitable. This gave rise to the plantation system in the late 1780s and the introduction of slavery. The institution of indentured servants never took hold in the north where the upkeep of slaves was not feasible since the growing season was so short, and it was unprofitable to maintain the slaves during unproductive times of the year.

The history of military warfare is replete with countless ways the weather factor has played a role in determining the outcome of battles and campaigns and in planning timing and strategies, on land, sea, and in the air. As early as 1274 written records tell of Mongol invasions of Japan being thwarted by windstorms, which crushed ships against rocky coasts, causing sinkings and a retreat.

Then, and in 1281, a devastating hurricane sank most of the naval vessels of Kublai Khan, the Mongol emperor. The Japanese credited divine winds with saving their country from foreign aggression. Those highly revered winds were called kamikaze, also a name given to Japanese suicide bomber pilots during World War II who were willing to give their lives in defense of their nation, then thought to be a most honorable and cherished manner of death.

A similar event in 1588 stymied the great Spanish Armada that sailed into the English Channel in preparation for a land invasion of England. The 132 large clumsy Spanish vessels were no match for the smaller more maneuverable English fleet that managed to open fire from close range. The Spanish galleons suffered huge losses in men and ships. In their attempt to escape and head back to Spain, they were met by tremendous gale force winds that destroyed all but a few of the remaining ships. Consequently, Spanish sea power was broken and England was able to rule the high seas. The foundations of that country's greatness in trade and the start of a worldwide colonial empire began with that victory and lasted hundreds of years.

Hannibal of Carthage (247–183 B.C.) was considered to be the greatest general of ancient times, yet he too fell victim to the harsh elements. In his military campaign to conquer Rome in 218–219 B.C., he attempted to march his armies across the Alps. Despite his genius in planning, he failed to take into account the snow, ice, and mudslides. Those elements and steep slippery mountain slopes caused him to suffer staggering losses of men, elephants, horses, and equipment, frustrating his ambitious goal of military conquest of the Romans. Weather conditions influenced the results of many engagements highlighting the military history of the United States. In the War for Independence, a strong gale on March 5–6, 1716, caused the British general to call off a planned attack on General George Washington's troops stationed south of Boston and the evacuation of the city by the British. A thick fog helped Washington's forces to evacuate Long Island enroute to New York and avoid a trap of his soldiers.

The timing of Washington's surprise crossing of the Delaware River on December 26, 1776, against the Hessians at Trenton, New Jersey, had to be revised due to thick ice on the river, sleet, driving rain, snow, and bitter cold temperatures. The equally cold winter of 1777–1778 at the campsite in Valley Forge, Pennsylvania, brought misery and struggles with survival for Washington's ill-equipped continentals. And the hard winter of 1779–1780 forced Washington to take refuge from military action at his winter quarters in Morristown, New Jersey. It was said to be the severest winter in American history. Yet, in the Battle of Monmouth, New Jersey, on June 18, 1778, during a heat wave, with temperatures on the field at 96 degrees F, Washington's soldiers were too fatigued to pursue the withdrawing enemy despite the fact that they outfought the British on the field of battle. As many as 109 soldiers on both sides of the fray died from the heat.

Other weather elements contributed to battle outcomes, as well. In the 1777 Battle of Bennington, Vermont, a torrential overnight rainstorm delayed Hessian troops from engaging undermanned American soldiers for twenty-four hours, enabling reinforcements to arrive and leading to the rout of British forces. At Saratoga, New York, an early snowfall on October 21, 1777, held up the British general's retreat into Canada, which led to the surrender of his entire army to American rebels, a turning point in the Revolutionary War. A very strong squall struck the British position under General Cornwallis at Yorktown, Virginia, on October 17, 1781, dividing his troops and preventing the arrival of reinforcements. Two days later he surrendered, and the war ended.

During the War of 1812 a powerful tornado, with destructive winds, hit Washington, D.C., on August 25, 1814, right after it had been set on fire by occupying British soldiers. The double ordeal caused many deaths and much damage in the capital. The Battle of New Orleans was fought during fog on January 8, 1815. Neither of the opposing forces, British or American, had visible contact, causing considerable misfiring of weapons and artillery, leading to a stalemate after General Andrew Jackson's troops defeated the British, killing and wounding more than 2,000 of the enemy.

In the War between the States, time and time again weather influenced troop movements and battle results. Marches were delayed and slowed by mud from rainstorms and other forms of inclement weather. Wind played a role in directing artillery fire and in launching balloon observation flights. Extremes of temperature, both hot and cold, led to battle fatigue, depression, and desertions in the ranks of soldiers on both sides of the fighting. Dust often clogged cannons. And in a bizarre encounter between soldiers of the Confederacy and the North at Lookout Mountain, Chattanooga, Tennessee, a battle was fought above the clouds on September 19–20, 1863. Troops were able to move undetected to high ground under cloud cover and lowered visibility.

In World War II, with the advent of air power, a great emphasis was placed on weather forecasting. Flights had to be strategically planned for cloud cover en route to targets and properly timed for clear skies over bomb sites. Winds aloft were of paramount importance. The extremely long "Russian Winter" was a key factor in aiding Soviet soldiers in defeating Nazi armies at the infamous battle of Stalingrad. The lifting of an extended December fog enabled American soldiers to counterattack at the Battle of the Bulge. The U.S. Navy lost a number of aircraft, warships, and supply vessels with tragic loss of lives in the Pacific Ocean during typhoons and storms at sea. During World War II, the Japanese made use of the Prevailing Westerlies wind flow. They were able to load explosives with timing devices onto balloons and have then drift toward the west coast of the United States.

Considerable weather forecasting went into the strategic planning of the D-Day invasion of France in early June 1944 during World War II. The ground troops, navy, and air force each required specific weather essentials that could best enable them to carry out their respective missions. Clear skies, dry land conditions, and calm winds over the water were optimal needs. Military weather forecasters relied on past analogous weather records to advise as to the logistics and ideal time when to make the landings on the continent of Europe.

Military personnel stationed worldwide must now be acclimated to function under all kinds of weather conditions, be they stationed in the Arctic, the desert, the tropics, or on top of snow-laden mountain peaks. Members of the armed forces are deployed rapidly to trouble spots the world over and must be prepared for fighting in any type of climate. They are the ones that often make history in unfamiliar environments.

HISTORY

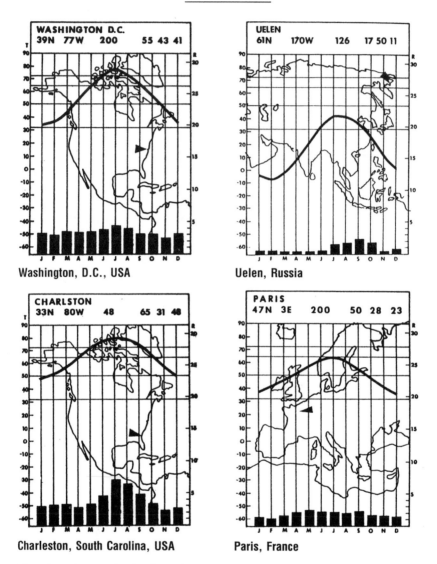

Washington, D.C., USA

Uelen, Russia

Charleston, South Carolina, USA

Paris, France

Historians frequently cite the unique role of weather in offering their notions as to meaningful underlying explanations for many of the events that have occurred in the past. They give descriptive accounts of how weather may have shaped the courses of historical events, military outcomes, cultures and institutions, explorations and settlements, national destinies, and even Biblical episodes. The places depicted in these climographs are some of the spots with unpredictable weather that may have affected history (see key in the Introduction).

SOURCES

Browning, Iben, and Nels B. Winkless, III. *Climate and the Affairs of Men*. New York: Harper's Magazine Press, 1975.

Clairborne, Robert. *Climate, Man and History*. New York: W. W. Norton, 1970.

Critchfield, Howard J. *General Climatology*. Englewood Cliffs, N.J.: Prentice-Hall, 1983.

Fagan, Brian. "Forgettable Weather." *New York Times*. July 30, 1999.

Farah, Mounir A., et al. *Global Insights*. New York: Glencoe, Macmillan/McGraw-Hill, 1994.

Lamb, H. H. *Climate, History and the Modern World* (2nd ed.). London: Routledge, 1995.

Laskin, David. *Braving the Elements: The Stormy History of American Weather*. New York: Doubleday, 1996.

Ludlum, David M. *The Weather Factor*. Boston: Houghton Mifflin, 1984.

———. "The Weather of Independence: The Battles of Trenton and Princeton." *Weatherwise*. November/December, 1998.

Neumann, J. "Great Historical Events that were Significantly Affected by the Weather." *Bulletin of the American Meteorological Society*, 56, Part 1. 1975.

Peterson, W. F. *Lincoln-Douglas: The Weather as Destiny*. Springfield, Ill.: C. C. Thomas, 1943.

Schneider, Stephen H., ed. *Encyclopedia of Climate and Weather*. New York: Oxford University Press, 1994.

Schneider, Stephen H., and Randi Louder. *The Coevolution of Climate and Life*. San Francisco: Sierra Club Books, 1984.

Spencer, J. E., and W. L. Thomas. *Introducing Cultural Geography*. New York: John Wiley and Sons, 1973.

Turner, Frederick Jackson. "The Significance of the Frontier in American History." In *The Annals of America*, vol. 2. Chicago: Encyclopaedia Britannica, 1963.

Wigley, T. M. L., M. J. Ingram, and G. Farmer, eds. *Climate and History: Studies in Past Climates and Their Impact on Man*. New York: Cambridge, 1981.

Wilder, Howard B., Robert P. Ludlum, and Harriet McCune Brown. *This Is America's Story*. Boston: Houghton Mifflin, 1966.

Various kinds of severe weather conditions lasting for long periods of time can bring dire hardships to those compelled to endure them. Their own survival may be at stake. As illustrated by this photograph, the harsh Dust Bowl storms of the late 1930s in the Great Plains region of the United States caused tens of thousands to leave the area and migrate to the western parts of America where it was hoped the pastures would be greener. Some even started their slow trek on foot. (Courtesy of U.S. Department of Agriculture.)

27

MIGRATION

From earliest times man has had a tendency to wander. It seems that migration is a normal form of human activity. The migrations of people have changed world history. Man moves from one place to another for various reasons. His motivation to migrate may stem from the over-population of his homeland, an enemy invasion, food shortage, perse-cution, tribal unrest, political suppression, the need to search for water and grasslands or new hunting fields, or to satisfy a curious adventurous nature. Perhaps survival was at stake. Then, of course, there was always the quest to improve his living standard by seeking new opportunities in a new land.

A direct and immediate reason for man to vacate an area was the sudden advent of natural catastrophes, such as an earthquake, fire, flood, drought, or volcanic eruption. When groundwater dried up, fields and pastures turned into dust bowls and desert; man had no alternative but to flee his native habitat. Often his life changed, sometimes radically, as the climate of his region changed, bringing hardship and an unwilling-ness to remain in such a trying environment.

In the United States people move to other regions for a variety of reasons including health, cost of living, job opportunities, and a change in lifestyles. But frequently older Americans seek to retire to states having milder weather. There has been a long trend for "northerners" to give up living in the cold, snowy tier of states to migrate to the warmer climes of Florida, Arizona, or California for a permanent resi-dence. Victims of repeated river floods often relinquished their homes and moved en masse to higher locations, sometimes to far away loca-tions. Tens of thousands of Plains farmers fled their farms for "greener pastures" in the western parts of America during the devastating Dust Bowl of the 1930s.

There remains a constant exodus from farm regions the world over due to crop failures brought on by the uncertainties of the weather, namely drought or excessive periods of rain. In the Sudan and other parts of the Savannah having seasonal rains in subtropical Africa, entire villages have abandoned their homesites during droughts, never to return. In Morocco in 1996 a yearlong drought, along with the drying up of village wells, caused thousands of villagers to resettle in that country's cities. With the influx of Moroccans leaving their farms and pasturelands, the cities became overcrowded with many homeless people.

Unfavorable climatic conditions intensified the ravages of the potato blight in Ireland in the 1840s. The sparse yield of that chief commodity drove more than a million people to leave their beloved country. Most Irish farmers and rural dwellers migrated to the large eastern cities of the United States, and few resumed their interest in working the land.

It is also interesting to note that migrants seek to move into areas where the climate and other physical features are somewhat similar to those of their native lands. Many of the Scandinavian immigrants to the United States, as a case in point, settled in Minnesota and Wisconsin. Many Italian immigrants who came from a Mediterranean climate, familiar with winter rains and summer droughts, tended to gravitate to a like climate in northern California, where their skills in vegetable growing and wineries, as well as fishing, could be put to good advantage. The wheat growing farmers of the Ukraine and Russia emigrated to the Canadian wheat belt of Manitoba where the two climates are almost identical.

Anthropologists and human geographers have uncovered a number of interesting findings about people who are transplanted to different climate regions. When people move from one climate area to another that has markedly different weather conditions, changes in the human makeup seem to take place, although the process is rather evolutionary. Men relocating from a cold or temperate climate to a hot region, though uncomfortable at the outset, are better able to withstand the heat and become acclimated more readily than women over a period of years. Some observers contend that women who move to the tropics begin to look paler than they appeared in their previous location, the effects of old age appear earlier, and vision efficiency deteriorates ten years earlier. Hot climates, studies indicate, may contribute to premature and greater memory loss.

When excessive alcohol is consumed in the tropical regions, the death rate tends to be much higher than in more temperate areas. It is thought that manual labor in the tropics is beneficial, particularly when sweating is very pronounced. A diet lower in protein and fat, but higher in carbohydrates is deemed better for people living in hot tropical climates. People who move from one extreme climate—like hot to cold—to the opposite tend to suffer greater incidences of neurological and psychological disorders.

Another kind of migration is very significant for American agriculture.

It is referred to as migrant labor, whereby farm workers come into the farm regions to plant and harvest crops during the growing season of the summer months. Most of the laborers come into the United States from Mexico or Central American countries and return to their native countries in the wintertime after their work contracts have expired and they are no longer needed in the fields.

As people are living longer and are able to enjoy fruitful lives in their golden years, it has become rather commonplace to have retirees migrate to warmer climates and take residence in adult communities throughout the south and southwest. And for those that may have emigrated to the United States during their younger productive years, many choose to return to their roots, in the countries they left in prior year, where they know the climate, culture, and environment and where they are comfortable in a familiar setting. This trend is especially prevalent for migrants who once left countries like Italy, Greece, Spain, and many Latin American nations where the climate is mild and more receptive. It is a kind of reverse migration.

The illegal migration of people can prove costly if their arduous and secretive attempts to enter another country are thwarted by unanticipated weather conditions en route. On April 2, 1999, a freak spring snowstorm trapped several dozen illegal immigrants in the wilderness mountain terrain east of San Diego, leaving at least eight people dead and many more missing. In 1999 more than 140 illegal immigrants died in two California counties during and immediately after crossing the border from Mexico. Many, not realizing how cold temperatures might be at night, were underdressed in light jackets and short-sleeved shirts. Those illegals that are rescued almost always need to be treated for hypothermia and exposure. Wetbacks, the Mexican laborers who illegally wade across the Rio Grande River between Texas and Mexico, have repeatedly suffered fatal drownings due to flash floods and high water encounters, especially during rainy times of the year.

For immigrants seeking to enter a new land illegally in overcrowded, nonseaworthy boats, their voyages can cost them their lives. There is a tragic record of Haitians, Cubans, and other refugees being lost in stormy seas. There have been sorrowful episodes of illegal migrants, often from the Orient, being let off large cargo ships onto small boats on wind-buffeted rough waters, within sight of the shoreline, only to drown trying to reach their destination. Many are picked up by the U.S. Coast Guard seamen or immigration authorities, only to be deported to their native land.

WEATHER DIFFERENCES OF URBAN AND RURAL AREAS

When people move to the cities from rural areas, most likely for economic opportunities, they are probably unaware that human activity can have an effect on weather. A study conducted by the National Science Foundation in 1922, entitled "Patterns and Perspectives in Environmental Science," found that there is a percentage difference in various meteorological factors in urban and rural areas. The disparities are attributed to man-made features found in cities, like the concrete, brick, and asphalt that absorb heat better than the trees and vegetation of the countryside.

Some differences can impact the health of city dwellers. Dust and smoke from automobiles and factories contaminate the atmosphere 1,000 times greater than the rural areas. Fog is appreciably greater in the cities. City temperatures are slightly higher as are thunderstorms, rainfall, and cloudiness. The investigation presented data that showed cities to have less solar radiation, visibility levels, wind speeds, and humidity. The snowfall differences were found to be less than persuasive. Some scientists feel that there is not enough evidence to support the findings, however, people who have lived in both rural and city environments will contend that there are recognizable differences based on their empirical experiences when they move from place to place.

MIGRATION

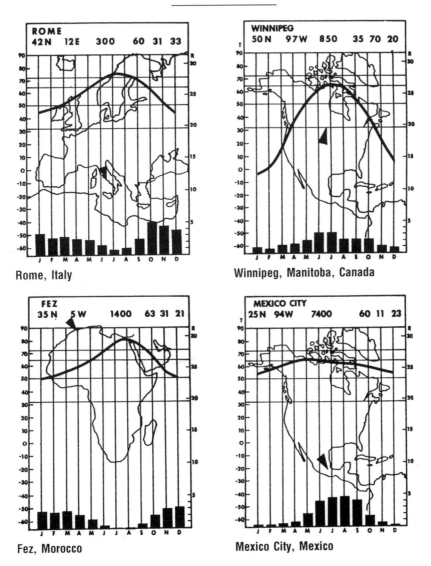

Rome, Italy

Winnipeg, Manitoba, Canada

Fez, Morocco

Mexico City, Mexico

Man has been known to show his inclination to vacate, wander, and migrate to another location for many reasons that may be directly or indirectly related to climate and weather. Sudden or sustained weather conditions, like a flood, drought, or another form of natural disaster, may cause him to uproot himself from his usual homesite and seek a more economically productive and secure environment. It is now a rather popular trend for people to relocate to regions having more hospitable climates where they may pursue the lifestyles of their choice. These climographs represent both areas from which people have migrated due to extreme weather conditions (Winnipeg and Fez) and areas to which people have migrated (Mexico City and Rome) (see key in the Introduction).

SOURCES

Carter, George F. *Man and the Land*. New York: Holt, Rinehart and Winston, 1975.

Dicken, Samuel N., and Forrest R. Pitts. *Introduction to Cultural Geography*. Lexington, Mass.: Xerox College Publishing, 1970.

Farah, Mounir A., et al. *Global Insights*. New York: Glencoe, Macmillan/McGraw-Hill, 1994.

Kurian, George Thomas, ed. *Geo-Data. The World Geographical Encyclopedia*. Detroit: Gale Research, 1989.

Newbill, Esko, and La Paglia. *Exploring World Cultures*. Lexington, Mass.: Ginn and Company, 1986.

Patterns and Perspectives in Environmental Science. National Science Foundation. Washington, D.C.: U.S. Government Printing Office, 1973.

Schneider, Stephen H., ed. *Encyclopedia of Climate and Weather*. New York: Oxford University Press, 1994.

Spencer, J. E., and W. L. Thomas. *Introducing Cultural Geography*. New York: John Wiley and Sons, 1973.

Statistical Abstract of the United States, 1998 (118th ed.). Washington, D.C.: U.S. Bureau of the Census, 1998.

Turner, Frederick Jackson. "The Significance of the Frontier in American History." In *The Annals of America*, vol. 2. Chicago: Encyclopaedia Britannica, 1963.

Wilder, Howard B., Robert P. Ludlum, and Harriet McCune Brown. *This Is America's Story*. Boston: Houghton Mifflin, 1966.

The World Almanac, 1999. Mahwah, N.J.: Primedia Reference, 1999.

The World Factbook, 1998. Washington, D.C.: Office of Public Affairs, Central Intelligence Agency, 1998.

Zwingle, Erla. "Morocco: North Africa's Timeless Mosaic." *National Geographic*. October 1996.

28

SHELTER

In addition to the need for food and clothing it is essential that man find or build a shelter to shield or protect himself from weather elements. This has always been a prime requirement for survival. Since climate and weather conditions vary from place to place, man has learned to design or build shelters that best suit local necessities. He also had to take into consideration the availability of nearby building materials and ways to protect himself from animals and enemies.

Generally in cold regions of the world, shelters had to be built to keep out snow, cold, and freezing winds. In hot, dry areas man was concerned with a burning sun and scorching winds. In the wet, hot lands the inhabitants of those climates had to construct shelters that would offer suitable refuge from heavy rains as well as the direct rays of a blazing sun. In more temperate climates man had to figure out ways to build shelters that could best adjust to the seasonal changes of both cold and hot temperatures, or dry and rainy months. Over the centuries improvements evolved in building dwellings; however, certain basic features and architectural designs tended to remain intact as efficient ways to contend with unique climates of a particular region.

In the tropics and other areas of hot temperatures at low latitudes, the nearly vertical sun makes its effect very intense and enervating for all humans. The air, which is hot and humid, increases body discomfort, necessitating shelters with wide roof overhangs to provide much-needed shade. Cone-shaped, thatched roof huts were and are common in equatorial areas. They gave protection from the sun and their pitched roof provided runoff from jungle rain. They are often assembled with parts of a matted wall that can be raised like curtains to allow the entry of outside air. Many are set on bamboo stilts or piles to allow air circulation beneath the floor in order to prevent interior dampness. This practice is prominent in Southeast Asia. Some huts are open-walled to get the

A condition called permafrost, or permanently frozen ground, occurs in almost half of Russia, and just about all of Siberia. The soil freezes to a great depth, sometimes up to a mile deep. During the short summer period when the ice thaws beneath the surface the foundations of homes, like the one in this picture, can sink as much as six feet (note height of the windows) over the years. (Courtesy of the author.)

greatest possible ventilation inside the dwelling. Some tropical climate tribes build huts in trees because of heavy rains.

The tents of desert nomads have flaps to offer shade during the day and restrict cool air at night merely by raising or lowering the coverings. In dry desert lands, nomads in search of grasslands practice transhumance, the seasonal movement of livestock and persons tending herds to and from lowlands and highlands. They often live in moveable tents and yurts. Yurts are circular tents of felt or skins used in Mongolia and nearby lands and are useful in cold or hot temperatures.

In climates that are quite cold or have a lot of snow, especially in the mountains of western Europe, Swiss chalet-type houses are built. They have thick walls of stone or thick timber to keep the cold out, and very steep reinforced roofs with overhanging eaves to handle heavy snowfalls. Time-proven stucco and brick facings on buildings serve well for many kinds of weather throughout much of Europe, especially in the cities. There the mid- and high-rise apartment complexes have balconies, some often used for storage. The balconies are primarily used on warm, balmy days and nights for fresh air and relief from the heat.

People of the high latitudes built their homes of logs, rocks, and earth to offset long periods of biting cold temperatures. Flat-topped houses abound in the American southwest, northern Africa, and the Middle-East, where it is usually hot and dry and where on very uncomfortable nights the occupants could find comfort sleeping on the roofs. Many of the sun-dried brick structures have small windows to keep out the intense heat and courtyards for outdoor cooking and relaxation. The typical feature of the Mediterranean region home is the open court called a patio where much family living is centered for open air activities. Mexicans build cool homes out of clay with overhang roofs for shade and have open breezy courtyards in which to relax. This type of home is also found on large estates and farms called haciendas.

The pueblo was a type of communal village built by certain Indians of the southwestern United States and parts of Latin America. Their dwellings consisted of a series of flat-roofed houses of stone or adobe and sun-dried brick arranged in terraces. Some were partially submerged underground so that it would be snug and warm during the winter and protected against the fierce plateau winds and moisture.

Most present-day Eskimos have never seen an igloo, or snow house, which today are rather rare in the Arctic region. Where they are built, they tend to be efficient. They can be constructed within an hour and then abandoned when it is no longer needed. Made from blocks of snow, they usually feature banks of snow at the base, a long entrance tunnel often dug beneath the ground, an elevated interior platform, and arched formations forming a dome or roof. Strangely, the colder the temperature outside, the warmer it can be inside. With a cold exterior temperature of −50 degrees F, you can have a temperature of 40–50 degrees F inside the igloo.

In the United States the effect of weather on homes can be both direct and indirect. Some homes are built with features meant to prevent storm damage. Sump pumps are often used for removing water to prevent basements from flooding. Special wind-resistant windows can be installed in homes and apartments located in areas prone to tornadoes, destructive coastal storms, hurricanes, and high velocity wind. Architects are now designing weatherproof homes that are built to withstand storms. They are being made of sturdier building materials, using better-secured roofing shingles, using pitched roofs to shed water and provide good air circulation, and utilizing wrap-around decks to capture and shield the wind. Some newly constructed homes are currently being built with reinforced interior bunkerlike rooms to withstand violent home-demolishing windstorms. In especially precarious locations near shore-lines, local building codes and insurance companies often require homes to be set on pilings twelve to twenty-five feet deep.

When selecting the site of a home, a number of climatic aspects need to be considered. Wind is very important. Free flowing air helps lessen humidity in a home. Determining prevailing local winds in advance, especially on a farm, can minimize the effect of oppressive farm animal odors, dust, and pollen carried toward a house. The location of a residence should also be planned for sun exposure and heating-cooling advantages at different times of the day and at different seasons of the year. This is becoming increasingly significant where homes are being equipped with solar heating panels. A place where frequent fog occurs can cause moisture damage to a home. Air pollution can also take a toll on the paint of a building's exterior, so too with frequent hailstorms, which can cause expensive damage. Also lightning rods and other protections may need to be a major consideration in house design.

People sometimes are compelled to select home sites for reasons beyond the climate factor, which are often personal, including availability and economics. Subsequently, the unforeseen disadvantages of the weather may manifest itself and cause dissatisfaction with their habitats. This is a real problem for some dwellers of damp caves dug in the hillsides of southern Spain, or in Kanduvan, Iran, where people in that area have lived in caves for more than 1,300 years.

In the former Soviet Union, during World War II, tens of thousands of rural timber homes and apartment blocks were destroyed by the invading Nazi armies. New precast apartment houses needed to be built immediately at war's end in 1946. The rush to construction along with the inferior materials, shoddy workmanship, and very harsh cold winters led to deterioration of the apartments and many are now in disrepair. In Siberia where buildings are built to keep the heat in and cold out many buildings have windows with three panes of glass to buffer the cold. Walls are twice as thick as those used in construction elsewhere in Russia. Entrances to buildings require passage through two or three doors. Steam heat is usually piped in from generating plants in another part of the city. Special building methods are used to prevent the foundations of individual homes from sinking and tilting when heated buildings are erected over frozen ground, which is called permafrost. The substratum is permanently frozen for a mile six feet below the surface, but it becomes a quagmire each summer during a time of thaw. Piles are driven into the frozen earth allowing a four-foot cushion of air to circulate, keeping the structure out of direct contact with the ground. In the Siberian taiga it is not uncommon to view an older house submerged into the ground with the first floor situated at the surface level of the earth, caused by years of slow gradual sinking each summertime.

SHELTER

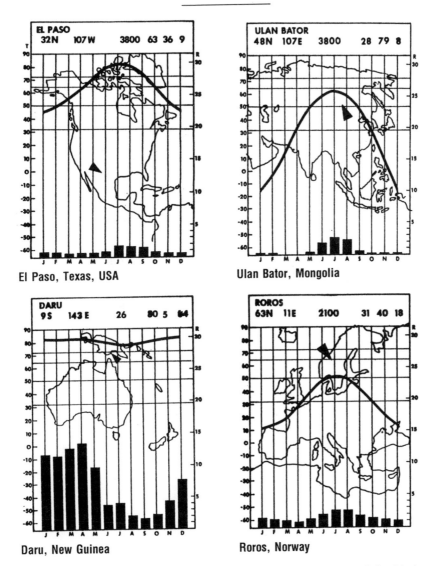

EL PASO
32N 107W 3800 63 36 9

El Paso, Texas, USA

ULAN BATOR
48N 107E 3800 28 79 8

Ulan Bator, Mongolia

DARU
9S 143E 26 80 5 64

Daru, New Guinea

ROROS
63N 11E 2100 31 40 18

Roros, Norway

The climate of a particular region may be a prime determiner of the kind and availability of specific building materials found in that region. That factor, along with the area's usual weather, be it temperature, rainfall, sun exposure, wind, snow, dampness, or a proven pattern of storm threats, may enter into the design, site selection, and construction of shelters. Today's architects still may replicate past styles and features in their new home renderings based on models that seem best suited to contend with and withstand the local elements over the years ahead. The climographs portray areas where extreme weather conditions have necessitated specific building styles (see key in the Introduction).

SOURCES

Aronin, J. E. *Climate and Architecture*. New York: Reinhold, 1953.

Ayoade, J. O. *Introduction to the Climatology for the Tropics*. New York: John Wiley and Sons, 1983.

Bragg, Rick. "Storm Over South Florida Building Codes." *New York Times*. May 27, 1999.

Brown, Patricia Leigh. "What the Hurricane Told the Architect." *New York Times*. July 24, 1997.

Carter, George F. *Man and the Land*. New York: Holt, Rinehart and Winston, 1975.

Dicken, Samuel N., and Forrest R. Pitts. *Introduction to Cultural Geography*. Lexington, Mass.: Xerox College Publishing, 1970.

Evans, M. *Housing, Climate, and Comfort*. London: Architectural Press, 1980.

Hecht, Esther. "Underground Living." *Jerusalem Post*. June 13, 1998.

Huriash, Lisa J. "New Home Built to Outlast Storms." *Sun-Sentinel* (Ft. Lauderdale, Fla.). January 8, 1998.

Ludlum, David M. *The American Weather Book*. Boston: Houghton Mifflin, 1982.

Resnick, Abraham. *Siberia and the Soviet Far East: Endless Frontiers*. Moscow: Novosti Press Agency Publishing House, 1983.

Schneider, Stephen H., and Randi Louder. *The Coevolution of Climate and Life*. San Francisco: Sierra Club Books, 1984.

Scioline, Elaine. "Cave Life Is Cozy, but the TV Reception Is Awful." *New York Times*. June 29, 1998.

Spencer, J. E., and W. L. Thomas. *Introducing Cultural Geography*. New York: John Wiley and Sons, 1973.

29

SPORTS AND RECREATION

Throughout the years an array of news stories reveal a close correlation between the weather and outcomes of many kinds of sporting events and recreational pursuits. Examples such as Rains Postpone Wimbledon Tennis Matches, Hurricane George Causes Cancellation of Miami–UCLA Football Game, Little League Baseball Finals Rescheduled Due to Wet Fields, Gale Force Winds Halt Golf Tournament, Early Thaw Melts Snow on Ski Slopes, Yanks–Orioles Game Rained Out, or Cold Snap Ends Swim Meet give clear indications of how the world of sports is affected by local weather conditions. Similar plans for family outings such as picnics, days at the beach, camping trips, sightseeing journeys, and fishing and boating excursions also may need to be changed or altered when the elements "act up." And most disheartening is the need to cancel reservations at vacation resorts and hotels when conditions appear to be unfavorable.

Seasonal temperature characteristics have a lot to do with the impetus of an early start and the evolution of certain sport activities of a particular region. Baseball, considered the national pastime, starts in the early spring and culminates when the World Series is over in the fall. The Major League players are fondly referred to as "the boys of summer." Tennis, swimming and diving, surfing, golf, beach volleyball, and track and field events are generally considered to be warm weather sports. Football, on the other hand, is basically played during the fall months. The arrival of snow and ice during the winter foster such outdoor activities as ice hockey, skiing, figure skating, bobsledding, and for the very courageous, ice climbing. For the exceptionally hale and hearty, able to withstand the biting winds and sub-freezing temperatures of frozen lakes and rivers, there are the exciting challenges of iceboating and ice fishing. Soccer, a universal and ever popular game, tends to know no season and is played seemingly rain or shine.

Beyond the United States and Canada, it seems that baseball has

Beach resorts with their "fun in the sun" offerings are worldwide weather friendly attractions for tourists and vacationers. Onshore seabreezes occur when the cooler, heavier ocean air moves inland to replace the lighter, rising air of the adjacent land. (Courtesy of National Oceanic and Atmospheric Administration.)

caught on in popularity in Japan and especially in Central America and the Caribbean Sea nations. The many warm days and nights in those areas afford athletes the opportunity to play lots of competitive year-round baseball games and allows them to hone their fielding and batting skills. Perhaps that is why there are now so many Latin American baseball players—estimated to comprise about 20 percent of the rosters for Major League teams. And Japanese ballplayers have also come into the majors. The ongoing warm temperatures serve as an incentive for professional baseball players to spend their off-season winters playing ball in the tropics in order to stay in "shape" and improve on their abilities. When spring training time arrives in March, most baseball players report to their club's facilities in Florida, Arizona, or California, where "baseball weather" seems to arrive a month or so early.

The athletes that participate in winter sports and have the ability to enter events at the Winter Olympic Games most often hail from native lands that have numerous lakes, ponds, canals, or water bodies that freeze over during so-called long winters. This enables them to become proficient at skating and ice hockey after months of practicing. Those countries that have Alpine slopes, where snow abounds throughout the winter, have an advantage in turning out excellent skiers and bobsledders.

Year in and year out, as demonstrated in Olympic competition, it seems that the best skiers invariably come from the snowfields of Europe and North America. In the high Alps Mountains of Switzerland no average daily temperature rises above freezing for six months.

In Moscow, Russia, and Quebec, Canada, the maximum average temperatures for four months of winter are below freezing. Spitzbergen, Norway, experiences eight months of average daily temperatures considerably below 32 degrees F. Each of these countries contributes fine specialized athletes to the winter Olympics. Canadians make up the largest percentage of National Hockey League players. Of late, a number of Russian hockey stars have joined their ranks as well.

The world's top surfers tend to hone their skills along the warm weather coasts of Hawaii, Australia, and Southern California where the surf is often high, driven by optimum wave-making winds.

Since most sporting events take place outdoors they are subject to the possibility of disruption, delay, postponement, or cancellation due to severe weather elements like rain, snow, wind, or cold. When there is a forecast of bad weather prior to a game, many fans are reluctant to watch the event in an open stadium or uncovered stands. When that happens the "no-shows" decrease attendance and ticket sales that go to support the teams. And if the sudden weather change brings discomfort to those in attendance during a game, many may decide to leave their seats and exit the stadium before the game ends. That is why many professional baseball or football games now take place in one of the ten domed indoor stadiums throughout United States and Canadian cities. The cavernous Louisiana Superdome in New Orleans is one such sporting facility. It is ironic that as an ominous hurricane threatened that city on September 27, 1998, thousands of fearful people took shelter in the dome. Even at the collegiate level, there is a large indoor stadium on the Syracuse University campus.

Over the years the quality of play has often been diminished by the results of bad weather on the playing field. That is very discouraging for player and fan alike. Games have been played on mud-soaked fields that resembled a lake or swamp, during blizzards when players were barely discernible, in fog conditions, or while very high relative humidity readings turned the field into a steambath for all. In some instances the players' benches have been augmented by moisture spray fans during hot unbearable temperatures, or by powerful heaters during subzero cold waves. On those days die-hard fans are at the stadium and the less committed are at home enjoying the game on the television sets in air conditioned or temperature-controlled rooms.

For the Boston and New York City Marathon runs in late April and late October, the temperatures and wind factors, though important for the runners, seem to be less crucial for the tens of thousands of race watchers who line the miles of sidewalks happily cheering their favorites to persevere and stay the course.

For the longest time it was generally assumed that winter was the dormant season for the normally sports-active. Many did, indeed, hibernate during the colder months. But now there is a growing trend for people, particularly the younger more hardy set, to become engaged in a broad array of winter sports, many of which take great physical stamina and endless courage. About fifty amateur mountain climbers die in a typical year climbing in the French part of the Mount Blanc massif. Alpine ice climbing of near vertical slopes with the use of an ice axe, an aluminum twelve-inch-wide ladder, ropes, metal clamps, and spiked boots is the ultimate sports challenge. It is incredibly daunting when the climber is 18,000 feet up, in thin air, battling an 8 degree F temperature, high winds, deep snow, frostbite, fatigue, altitude sickness, and falling chunks of ice.

For those who have aspired to get to the roof of the world at Mount Everest at 29,028 feet, the world's highest mountain has claimed more than 150 lives, including five in 1996 and eight in 1997. Nevertheless, ice and snow climbing is still popular on seven continents.

Snow and ice enthusiasts seem to have a curious love affair connecting sports, competition, festivals, and "King Winter." Hunters ardently await the winter hunting season to forage the woods for deer and other game. At Nome, Alaska, a great test of human and husky endurance takes place with the annual sled-dog race, known as the Mardi Gras of the Arctic. It is a highly promoted affair. All across Russia each winter very hardy bathers and swimmers take dips into icy lakes and rivers on weekends when the air and water temperatures are well below freezing. They call themselves walruses. Those who watch from the shore may call them lummoxes, implying stupidity.

In Hindeloopen, Netherlands, a thirteenth-century trading port, "fancy skating" by large groups on the frozen canals is a culture of the region. Thousands turn out to watch the routines of competitive teams during their festivals. Dutch towns also compete with each other with 200-kilometer (124 miles) skating marathons that can draw 16,000 participants and a half-million onlookers.

THE WEATHER-CONSCIOUS TRAVEL AND RESORT INDUSTRY

Throughout the world there are many wonderfully picturesque sites that attract the interests of curious travelers eager to experience and film what Mother Nature has to offer. They include such weather-formed geographic phenomenon as glaciers, waterfalls, river rapids, lakes, everglades, wind-driven land formations, colorful canyons, sand dunes, and other unique physical features.

The huge travel and resort industries are extremely weather-conscious. Their brochures and advertisements constantly extol the ideal climates that their attractions profess to offer the prospective sports-minded vacationers. Unfortunately for the travel agents, hotel managers, and Chambers of Commerce, when the weather turns out to be something less than anticipated by the tourists or guests, their public relations suffer. But, with the climatic laws of average prevailing, most visitors usually return for a second chance at having sporting fun—in the sun or snow.

THE EFFECT OF WEATHER ON LITERATURE AND THE ARTS

For many people their recreational preferences center on such pursuits as reading, dance, music, or drama. In each of these endeavors climate and meteorological imagery are frequently used as a motif, theme, or underlying backdrop to the storyline of the literature or lyrics of a song. Writers and other creative artists often resort to the use of weather terms as metaphors in order to provide the reader or audience with meaningful insights or provocative grounds for making analogies and formulating messages based on the script. It is not uncommon to find that authors rely on such implied, descriptive words as foggy, misty, cloudy, windy, or stormy to set a tone for a particular scene.

The literature is replete with climatic references and subtleties by some of the most outstanding literary figures of all time. Wadsworth, Shakespeare, and Milton allude to seasonal changes in their comparisons to the passing of time and the transitory character of human existence. Shelley in his "Ode to the West Wind" also writes about the weather's changing movement. He advances the notion that "we are as clouds" that are soon lost forever. The Puritans, obsessed with religious explanations, warn readers about the meaning of thunder and lightning in the context of sin. Jonathan Edwards and other Puritan writers reveal that the wrath of God manifests itself in the form of hail, whirlwinds, black clouds, and dreadful storms. Herman Melville's *Moby Dick* also characterizes a storm at sea. Paul Colan describes the blurring force of a blizzard. Emerson and Hawthorne provide word pictures of snow landscapes. In Coleridge's *Rime of the Ancient Mariner* the title correlates pun and poem.

In the history of American theater and filmmaking the weather factor has not only played a significant role in affecting the plots of the story, but some meteorological element can often become part of the title. Examples of this are illustrated by such artistic works as *The Four Seasons*, *Fog*, *The Rainmaker*, *The Lion in Winter*, *Body Heat*, and *Flash of Lightning*. Such classic songs performed on stage and screen, namely "Summertime," from Gershwin's *Porgy and Bess*, "Stormy Weather," "Singing in the Rain," and Irving Berlin's "White Christmas," have entertained international audiences for decades. Popular songs that allude to the weather include such hits as "On The Sunny Side of the Street," "Raindrops Keep Fallin' on My Head," and "You Are the Sunshine of My Life."

Many kinds of informal weather terms have been put to use by authors and creative artists as coined works and expressions when drawing analogies, illustrating themes, or describing emotions. Some pithy examples now in popular use are: "a clouded past," "chill out," "warmed to the occasion," "snow job," "thunderous applause," "cool," "all wet," "warmed-up," "hot potato," "bag of wind," "luke-warm," "cold personality," and "frozen in his tracks."

Some authors, using their powers of imaginative expression, were able to use various weather elements to stress their inner feelings about the issues of their time. John Steinbeck's *Grapes of Wrath* focused on the drought to call attention to the disadvantaged lives of poor farmers. Walt Whitman ties the events of the Civil War and its great battle to a great storm. In Harriet Beecher Stowe's *Uncle Tom's Cabin* the reader empathizes with the runaway slave Eliza escaping from her pursuers by running across blocks of ice. Washington Irving, Nathaniel Hawthorne, and Henry Thoreau, in their weather stories, link weather changes to seasonal changes in landscapes with their vivid selections of word descriptions.

Drama and dance performances in Egypt, Greece, India, Japan, and Africa continue to use weather elements for ancient-type dramatic effects in their seasonal festivals. There are centuries-old spring planting ceremonies and presentations paying homage to the sun goddess. In many Native American cultures, dance rituals, such as the Navajo water ceremony, have been performed over the years.

In some William Shakespeare's plays (i.e., *The Winter Tale* and in *A Midsummer Night's Dream*, as well as in *King Lear*), there are connecting themes of human emotions and weather conditions. A number of his stories, as well as the works of other authors, like Lord Byron, inspired choreographers and composers to produce ballets and operas. The performances were almost always highlighted by dramatic flashes of lightning and crashing thunder as they and other elements served to underscore various human emotional aspects and responses. In the ever-popular *The Nutcracker*, the magical Christmastime ballet's stage setting is made realistic and exciting by the effects of artificial snow and fog that serve to enhance the dynamic colorful costumed dancers and its delightful music.

SPORTS AND RECREATION

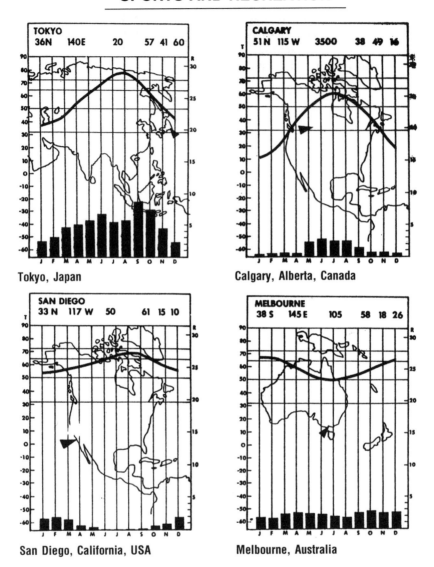

Tokyo, Japan

Calgary, Alberta, Canada

San Diego, California, USA

Melbourne, Australia

Athletes that have the opportunity to play a sport in regions having ideal weather conducive for that game may have climate advantages over other athletes limited in their play because of unsuitable weather or a shorter season for participation where they live. The kind of typical weather most likely found in an area often lends itself to the specialization and production of a greater number of athletes that excel in that particular sport. Good weather that affords the frequency of participation helps prove the adage that "practice makes perfect." The travel and resort industry is extremely weather-conscious in promoting their offerings to prospective clients. These climographs depict areas suitable to both warm-weather sports like baseball as well as cold-weather sports like ice skating (see key in the Introduction).

SOURCES

Banham, M., ed. *The Cambridge Guide to World Theatre*. New York: Cambridge University Press, 1988.

Bordman, G. M. *The Oxford Companion to American Theatre* (2nd ed.). New York: Oxford University Press, 1992.

Brown, Gerry, and Michael Morrison, eds. *The 1999 Information Please Sports Almanac*. Boston: ESPN Books, 1999.

Friend, Tim. "Ice Climbing, and Clinging, 18,000 Feet Up." *USA Today*. May 19, 1998.

Goldberg, Carey. "Deep in the Doldrums of Alaska's Winter, a Mardi Gras for Mushers." *New York Times*. March 13, 1997.

Lockhart, G. *The Weather Companion: An Album of Meteorological History, Science, Legend and Folklore*. New York: Wiley, 1988.

Schneider, Howard. "Canadians Have Curious Love Affair with Winter." *Sun-Sentinel* (Ft. Lauderdale, Fla.). March 8, 1997.

Schneider, Stephen H., ed. *Encyclopedia of Climate and Weather*. New York: Oxford University Press, 1994.

Smalling, R. J. *The Sport Americana Address List*. Cleveland: Edgewater Book, 1994.

Sparano, Vin T., ed. *Complete Outdoors Encyclopedia*. New York: St. Martin's Press, 1998.

Specter, Michael. "Cold War of the Soul: Siberian Ice Fishing." *New York Times*. February 10, 1998.

Wallechinsky, David. *The Complete Book of the Olympics*. New York: Viking, 1998.

"Weather Goes from Bad to Worse, and So Does the Ski Schedule." *New York Times*. February 15, 1998.

Whitney, Craig. "Alpine Thrills Attract an Untiring Visitor: Death." *New York Times*. August 17, 1997.

The World Almanac and Book of Facts, 1996. Mahwah, N.J.: Primedia Reference, 1999.

Buffalo, New York, is a city well known for its severe winter blizzards. Record snowfall and low temperatures paralyzed the city in February 1977, affecting transportation throughout the city. Here Red Cross volunteers on snowmobiles look to rescue persons that may be trapped in their car. (Courtesy of American Red Cross.)

30

TRANSPORTATION

During earlier times as travelers began to venture from their local towns, there was great concern that their wagons might be halted by road robbers. Travel on the first American roads, though unsafe, was also uncomfortable and extremely slow. The dirt surfaces were dusty in summer and muddy in winter. Weather conditions frequently closed the roads and delayed trips for days on end. Even the arrival of privately built toll roads or turnpikes in the late 1700s failed to improve the state of colonial land transportation. Weather was always a prime factor then in how or when people and goods could be transported.

Today's great multilane interstate highways and high-speed turnpikes are much safer and more comfortable for travel, especially during times of hazardous road conditions due to weather. Highways are quickly sanded during icy condition, and snow is removed almost as fast as it falls. Work crews are placed on advance alert when winter storms are forecast. Motorists are forewarned about the kind of weather they will encounter en route by large well-lighted signs telling about snow, ice, blowing sand, or fog dangers.

Trailers are most often banned from entering turnpikes when high winds are reported. Speeds of vehicles are reduced when weather conditions pose a threat to safe driving. Highway police on patrol are always ready to set warning flares on the roads to induce drivers to slow down in areas where smoke or haze might reduce visibility. In the interest of safety, permanently mounted signs such as "Road May Be Flooded," "Caution: Crosswinds on Bridge," or "Bridge Freezes Before Roadway" are now posted along interstate routes. Also, coastal areas now erect "Evacuation Route" signs in anticipation of possible storms.

Large snow accumulations can close a feeder road to a major highway. The most treacherous road hazard of all occurs when freezing rain causes layers of "black ice" to form suddenly at unforeseen parts of a roadway. Unsuspecting drivers reaching the icy spot are unable to stop and most assuredly lose control of their vehicles. This frequently leads to disastrous multicar pileups.

According to statistics made available by the National Highway Traffic Safety Administration, the weather factor is significant in causing crashes of vehicles on American roads. In 1995 weather conditions caused about 18 percent of all accidents. Rain was blamed for an estimated 13 percent, snow/sleet 3.5 percent, and other weather conditions approximately 1.5 percent. The total fatal crashes that year amounted to 37,221. Of those, 3,165 were attributed to rain, and 907 to snow/sleet. In addition to human losses due to death and injury crashes, there are staggering dollar damages to property brought on by bad weather. There are also many weather-related problems that can delay and affect the routines of freight movement and mail deliveries, commuting to work, school bus arrivals, and the timing of business and social meetings. When workers in hospitals, medical offices, police and firefighters, or emergency personnel on ambulances cannot report for duty due to severe weather conditions, the results can be quite critical.

The American Automobile Association recommends that all operators of motor vehicles be prepared for extreme, drastic weather that can lead to a car breakdown. Checking anti-freeze, hoses, and winterizing of the engine are essential precautions. Radiators and tires need to be examined. Some roads require chains or snow tires during months when snow impacts road conditions. Drivers who regularly encounter varied driving conditions due to weather might consider purchasing front or four-wheel-drive vehicles that can better maneuver in mud, snow, or ice.

ANIMALS SUITED TO PROVIDE TRANSPORTATION

In many places, man still has to rely upon animals to provide transportation. Sometimes primitive kinds of transportation and newer mechanical means are used in the same region. Animals that carry loads on their back are called beasts of burden. The most common beasts of burden are horses, burros, mules, and oxen. Selective types of animals can live and work in a variety of climates for which they are well suited. Camels for example, can walk on hot desert sands without water for several days. The llamas of mountainous South America can live and work at high altitudes in cold temperatures. The long hair of yaks makes them good carriers in the cold regions of Tibet. Reindeer can also stand the cold. In the monsoon climate of India, elephants perform well and can carry heavy loads. The strong water buffalo of South Asia, Malaysia, and the Philippine Islands likes to wallow in mud and wet rice paddies and is used as a draft animal. Burros have been used over the years in the southwestern United States where they are efficient on steep terrain in warm, dry climates. In the hot countries of the world, the ox is relied on for pulling big loads. And in the cold and snow of the arctic, dog sleds, though now relatively rare, are used for special occasions where transportation of any kind is very hard to come by.

TRANSPORTATION DANGERS AND DILEMMAS DURING TIMES OF DIRE WEATHER

During the coldest part of the tough winters of the Canadian Northwest Territories, almost all big lakes and small ponds in a 400-mile region freeze over solid four feet deep. That is when the "ice highway" opens for driving mammoth twenty-five-foot-wide, 150,000-pound dump trucks over what the truckers call the "blue highway" to and from Arctic area diamond and ore mining companies. Driving the huge eighteen-speed diesel rig along this stretch can be as boring as it is dangerous. Many a driver has plunged through a "safe" ice lake. For miles on end, the landscape is white on white snow. About the only distraction may be a wait of an hour or more while a herd of caribou crosses the make-shift roadway.

The harsh winter climate of Siberia and much of Russia's northern tier offers the best example of how forbidding weather can affect transportation. Buses and trucks need to have their engines running at frequent intervals throughout the night in order to prevent their batteries from dying. Private automobiles are wisely stored in public garages for the coldest months. Those vehicles that are needed in winter have special heavy-duty batteries installed; kerosene is mixed with engine oil and soft rubber tires replace normal tires. Every vehicle has a double windshield with a sealed air space to prevent frosting and fogging.

Roads in Siberia are often very hazardous and extremely dangerous to drive. They become clogged by winter frosts, raging snowstorms, or armies of summer mosquitoes that can obscure drivers' vision for miles. Weather-beaten roads have a short duration of usability after they are built. They suffer considerable heaving and disintegration due to the underlying permafrost and surface melting. Some even sink and vanish in the summer. On certain rivers, after the ice runs deep during the winter months, trucks actually roll along the frozen river highways, a very dangerous practice resorted to by more daring drivers.

The Siberian driver, as well as those in other remote cold regions, must be certain that they have adequate amounts of fuel when traveling through isolated forests and sparsely settled territories. Running out of gas miles from "nowhere" can easily be fatal. Many have frozen to death under such circumstances. With winter surface travel in Siberia and other Arctic regions being so treacherous, alternative forms of transportation take hold. Ice-breaking vessels are used to unclog ice jams on rivers and other water bodies so that people and goods can be transported by boat.

Some rivers in Siberia are ice-free only two or three weeks per year. Extensive use of aircraft to fly over frozen and springtime flooding of roadways is greatly relied upon now more than ever. But all too often extremely bad winter weather closes airports or diverts and delays scheduled flights, sometimes for days. This can cause massive backups of stranded passengers and cargo at terminals before the weather clears enough for flights to resume. Hundreds of anxious and frustrated travelers are sometimes required to sleep on the airport terminal floors, taxing the limited facilities beyond imagination. As a result of Siberia's many weather obstacles, supplies of food, goods, and services often become scarce and very expensive.

To offset Russia's transportation dilemmas when dire weather occurs, the country relies on a very efficient network of railways, which can also be found throughout most of Europe as well. Trains rarely are impeded from meeting their time schedules by storms or other weather-related elements. The Trans-Siberian Railroad is a good example of this. It spans two continents from Moscow to the Pacific Ocean, more than 4,000 miles, and its seven-day journey is almost always guaranteed to arrive on time, no matter what weather is encountered along its route. European countries count on the use of their canals and rivers as important transportation arteries, except during the freezing months of winter when all navigation ceases due to ice.

For pilots, flight engineers, air traffic controllers, dispatchers, airport operational personnel, weather forecasters, and ground crews, the meteorological factors involved in the flying of an aircraft is paramount for the safety of the crew and the comfort of the passengers at all times. Knowledge of weather conditions in flight and at airports that affect the success of a particular flight are often of paramount importance. Those aspects that affect flying are wind speed and direction, cloud ceilings, ground visibility such as fog, turbulence, lightning, hail, thunderstorms, snowstorms, downbursts of air, heavy rainfall, and ice on the plane's wings. For the 300,000 plus private pilots holding licenses in the United States, a thorough familiarity with weather could be a lifesaver since a large percentage of small aircraft crashes are weather-related. On the other hand, United States airline safety for scheduled commercial carriers between 1979 and 1994 averaged about three fatal accidents per year, which, considering the overall flights recorded, is rather remarkable.

In recent times commercial air travel has become increasingly safe in regard to the weather factor. State-of-the-art radar, accurate flight planning generated by excellent computerized weather forecasting, very reliable upper-air data collection, and up-to-the-minute in-flight and destination weather reports assist pilots in avoiding the worst frontal storms aloft as well as zones of turbulence. Sophisticated "automatic pilots" and computer guidance systems now enable pilots to fly and land huge airplanes with relative ease, even in the thick of the night during an uninviting storm or weather disturbance. The days of "flying by the seat of your pants," as the expression goes, are thankfully now over!

Despite the tremendous advances in aircraft safety features 70 percent of air traffic delays are attributed to the weather factor. At times passengers and planes can be "backed up" at airport terminals for hours, even when bad weather takes place hundreds of miles away. The guideline "it is better to be safe than sorry" proves that patience pays.

TRANSPORTATION

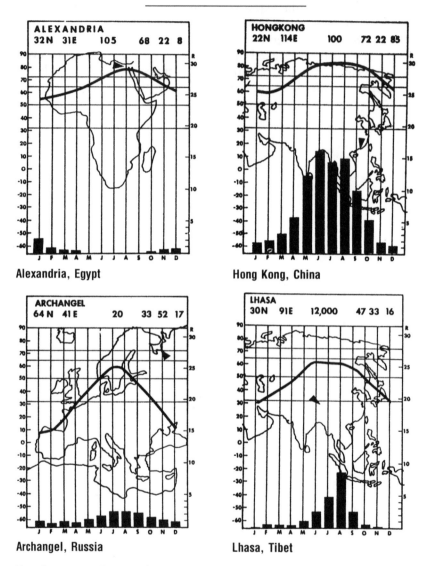

Alexandria, Egypt

Hong Kong, China

Archangel, Russia

Lhasa, Tibet

Despite great advances in applying advanced technological improvements and safeguards to all forms of today's transportation, the weather factor still persists as a formidable challenge to most kinds of transports, on land, sea, and in the air. Extremely unfavorable weather conditions can close roads, delay rail traffic, shut down airports, and cancel sailings, as often happens in the areas represented in these climographs (see key in the Introduction). Transportation officials and associations are constantly examining ways to reduce weather-related accidents, particularly those causing serious injuries and fatalities.

SOURCES

Boselly, S. E., III, et al. *Road Weather Information Systems, Vol. 2., Implementation Guide*. Washington, D.C.: National Research Council, Strategic Highway Research Publication, H-351, 1993.

Battan, Louis J. *Weather In Your Life*. San Francisco: W. H. Freeman, 1983.

Borenstein, Seth. "Ice Endangers 'Atlantis' Launch." *Orlando Sentinel*. January 11, 1997.

Crashes by Weather Condition. Washington, D.C.: U.S. Department of Transportation, National Highway Traffic Safety Administration, 1995.

Critchfield, Howard J. *General Climatology*. Englewood Cliffs, N.J.: Prentice-Hall, 1983.

Fabre, Maurice. *A History of Land Transportation* (Vol. 7.). New York: Hawthorne Books, 1963.

Farah, Mounir A., et al. *Global Insights*. New York: Glencoe, Macmillan/McGraw-Hill, 1994.

Fisher, James S. *Geography and Development: A World Regional Approach*. New York: Macmillan, 1992.

Greene, D. L., and P. J. Santini, eds. *Transportation and Global Climate Change*. Washington, D.C.: Department of Transportation, 1993.

Golaszewski, R. *Weather Briefing Use and Fatal Accidents*. Washington, D.C.: National Research Council, Transportation Research Board, No. 1158, 1988.

Oliver, John E., and Rhodes W. Fairbridge, eds. *The Encyclopedia of Climatology*. New York: Van Nostrand Rheinhold, 1987.

Schneider, Stephen H., ed. *Encyclopedia of Climate and Weather*. New York: Oxford University Press, 1994.

Spencer, J. E., and W. L. Thomas. *Introducing Cultural Geography*. New York: John Wiley and Sons, 1973.

Statistical Abstract of the United States 1998 (118th ed.). Washington, D.C.: U.S. Bureau of the Census, 1998.

The World Almanac, 1999. Mahwah, N.J.: Primedia Reference, 1999.

APPENDIX: DIRECTORY OF WEB SITE CONTACTS

WEATHER-RELATED GOVERNMENT AGENCIES

American Red Cross
http://www.crossnet.org

Avalanches Information
http://www.alaska.net/-nwsar

Climate Diagnostic Center
http://www.cdc.noaa.gov

Climate Monitoring and Diagnostics Laboratory
http://www.cmdl.noaa.gov

Climate Prediction Center
http://www.nnic.noaa.gov/cpc

Decoded Offshore Weather Data
http://www.ems.psu.edu/cgi-bin/wx/offshore.cgi

Earth Science Enterprise
http://www.hq.nasa.gov/office/mtpe

Environmental Information Services
e-mail: help@esdim.noaa.gov

Famine Early Warning System
http://www.info.usaid/gov/fews/fens.html

Federal Aviation Administration
http://www.faa.gov

Forecast Systems Laboratory
http://www.fsl.noaa.gov

Geophysical Fluid Dynamics Laboratory
http://www.gfdl.gov

Global Change Data and Information System
http://www.gdis.usdcrp.gov

Global Hydrology and Climate Center
http://www.ghec.msfc.nasa.gov

National Air Space Administration (NASA)
http://www.hq.nasa.gov

National Climate Center
http://www.noaa.gov

National Climate Data Center (NCDC)
http://www.ncdc.noaa.gov

National Environmental Satellite Data and Information Service
http://www.ns.noaa.gov/NESIS/NESDISHome.html

National Highway Traffic Safety Administration
http://www.nhtsa.dot.gov

National Hurricane Center
http://www.nhc.noaa.gov

National Oceanic and Atmospheric Administration
http://www.noaa.gov

National Severe Storms Laboratory
http://www.nssl.noaa.gov

National Transportation Safety Board
http://www.ntsb.gov

National Weather Service
http://www.nws.noaa.gov

National Weather Service Alaska Region
http://www.alaska.net/-nwsar

NESDIS Public Affairs
pviets@nesdis.noaa.gov

Tropical Prediction Center National Hurricane Center
http://www.nhc.noaa.gov

WIND and SEA
http://www.lib.noaa.gov/docswindandsea

MISCELLANEOUS U.S. GOVERNMENT AGENCIES

Census Bureau
http://www.census.gov

Department of Agriculture
http://www.usda.gov

Department of Commerce
http://www.doc.gov

Department of Energy
http://apollo.osti.gov

Department of Health and Human Services
http://www.os.dhhs.gov

Department of Housing and Urban Development
http://www.hud.gov

Department of Interior
http://www.doi.gov

Department of Transportation
http://www.dot.gov

Environmental Protection Agency
http://www.epa.gov

Federal Bureau of Investigation
http://www.fbi.gov

Federal Emergency Management Agency
http://www.fema.gov

Library of Congress
http://www.loc.gov

NASA
http://www.nasa.gov

National Institute of Health
http://www.nih.gov

NOAA NATIONAL DATA CENTERS

National Climatic Data Center
ordens@ncdc.noaa.gov

National Geographical Data Center
info@ngdc.noaa.gov

National Geographic Society
www.nationalgeographic.com

National Oceanographic Data Center
services@nodc.noaa.gov

RIVER FORECAST CENTERS

Alaska River Forecast Center
Anchorage, Alaska
To access: http://www.alaska.net/~akrfc

Arkansas-Red Basin River Forecast Center
Tulsa, Oklahoma
To access: http://www.info.abrfc.noaa.gov

California-Nevada River Forecast Center
Sacramento, California
To access: http://www.nimbo.wrh.noaa.gov/cnrfc

Colorado Basin River Forecast Center
Salt Lake City, Utah
To access: http://www.cbrfc.gov/home.html

Lower Mississippi River Forecast Center
Slidell, Louisiana
To access: http://www.srh.noaa.gov/ftproot/orn/html

Middle Atlantic River Forecast Center
State College, Pennsylvania
To access: http://www.crab.met.psu.edu

Missouri Basin River Forecast Center
Pleasant Hill, Missouri
To access: http://www.crh.noaa.gov/mbrfc/index.html

North Central River Forecast Center
Chanhassen, Minnesota
To access: http://www.crhnwscr.noaa.gov/ncrfc/welcome.html

Northeast River Forecast Center
Portland, Oregon
To access: http://www.tgsv5.nes.noaa.gov/er/nerfc

Ohio River Forecast Center
Wilmington, Ohio
To access: http://www.nwrfc.noaa.gov

Southeast River Forecast Center
Peachtree City, Georgia
To access: http://www.srh.noaa.gov/ftproot/atr/html/main_p.htm

West Gulf River Forecast Center
Fort Worth, Texas
To access: http://www.srh.noaa.gov/wgrfc

WEATHER WEB SITES

National Weather Service Home Page
http://www.nws.noaa.gov

National Weather Service Office of Meteorology
http://www.nws.noaa.gov/om/omhome

National Center for Environmental Prediction
http://www.ncep.noaa.gov

National Drought Mitigation Center
http://www.enso.unl.edu/ndmc

Office of Hydrology (flooding)
http://www.nws.noaa.gov/oh

Storm Prediction Center
http://ww.nssl.noaa.gov/~spc

Tropical Prediction Center
http://www.nhc.noaa.gov

Ultraviolet (UV) Index Format
http://www.nic.fb4.noaa.gov/products/stratosphere/uv_index

Weather Calculator
http://nwselp.epcc.edu/elp/wxcalc.html

Weather Net
http://cirrus.sprl.umich

The Weather Page
http://acro.harvard.edu/GA/weather.html

INDEX

About the Author

ABRAHAM RESNICK is a retired Professor at Jersey State College and former Director of the Instructional Materials Center at Rutgers University Graduate School of Education. He served as a weatherman in The United States Army Air Corps during World War II.